AF457821

aus der Reihe:

Innovationen mit Mikrowellen und Licht

Forschungsberichte aus dem Ferdinand-Braun-Institut, Leibniz-Institut für Höchstfrequenztechnik

Band 58

Dimitri Stoppel

Interconnection development for InP-HBT terahertz circuits

Herausgeber: Prof. Dr. Günther Tränkle, Prof. Dr.-Ing. Wolfgang Heinrich

Ferdinand-Braun-Institut
Leibniz-Institut
für Höchstfrequenztechnik (FBH)
Gustav-Kirchhoff-Straße 4
12489 Berlin

Tel. +49.30.6392-2600
Fax +49.30.6392-2602

E-Mail fbh@fbh-berlin.de
Web www.fbh-berlin.de

Innovations with Microwaves and Light

Research Reports from the Ferdinand-Braun-Institut, Leibniz-Institut für Höchstfrequenztechnik

Preface of the Editors

Research-based ideas, developments, and concepts are the basis of scientific progress and competitiveness, expanding human knowledge and being expressed technologically as inventions. The resulting innovative products and services eventually find their way into public life.

Accordingly, the *"Research Reports from the Ferdinand-Braun-Institut, Leibniz-Institut für Höchstfrequenztechnik"* series compile the institute's latest research and developments. We would like to make our results broadly accessible and to stimulate further discussions, not least to enable as many of our developments as possible to enhance everyday life.

This work investigates key aspects in indium phosphide process development, and the findings helped improving FBH's transferred-substrate process. The developments in this work make it possible for the first time to build a fully functional millimeter-wave radar with integrated terahertz components. This opens new paths for complex high-frequency systems based on indium phosphide heterobipolar transistors, thus paving the way for future terahertz projects and applications.

We wish you an informative and inspiring reading

Prof. Dr. Günther Tränkle
Director

Prof. Dr.-Ing. Wolfgang Heinrich
Deputy Director

The Ferdinand-Braun-Institut

The Ferdinand-Braun-Institut researches electronic and optical components, modules and systems based on compound semiconductors. These devices are key enablers that address the needs of today's society in fields like communications, energy, health and mobility. Specifically, FBH develops light sources from the visible to the ultra-violet spectral range: high-power diode lasers with excellent beam quality, UV light sources and hybrid laser systems. Applications range from medical technology, high-precision metrology and sensors to optical communications in space and integrated quantum technology. In the field of microwaves, FBH develops high-efficiency multi-functional power amplifiers and millimeter wave frontends targeting energy-efficient mobile communications as well as car safety systems. In addition, compact atmospheric microwave plasma sources that operate with economic low-voltage drivers are fabricated for use in a variety of applications, such as the treatment of skin diseases.

The FBH is a competence center for III-V compound semiconductors and has a strong international reputation. FBH competence covers the full range of capabilities, from design to fabrication to device characterization.

In close cooperation with industry, its research results lead to cutting-edge products. The institute also successfully turns innovative product ideas into spin-off companies. Thus, working in strategic partnerships with industry, FBH assures Germany's technological excellence in microwave and optoelectronic research.

Interconnection development for InP-HBT terahertz circuits

Dimitri Stoppel

Von der Fakultät IV - Elektrotechnik und Informatik -
der Technischen Universität Berlin

zur Erlangung des akademischen Grades
Doktor der Ingenieurwissenschaften, Dr.-Ing.

Promotionsausschuss:

Gutachter: Prof. Dr. Günther Tränkle
Gutachter: Prof. Nils Weimann, Ph.D.
(University Duisburg-Essen)
Gutachter: Prof. Dr. habil. Wolfgang Heinrich

Juli 2019

Bibliografische Information der Deutschen Nationalbibliothek
Die Deutsche Nationalbibliothek verzeichnet diese Publikation in der Deutschen Nationalbibliographie; detaillierte bibliographische Daten sind im Internet über http://dnb.d-nb.de abrufbar.
1. Aufl. - Göttingen: Cuvillier, 2020
Zugl.: (TU) Berlin, Univ., Diss., 2019

Nonnenstieg 8, 37075 Göttingen
Telefon: 0551-54724-0
Telefax: 0551-54724-21
www.cuvillier.de

1. Auflage, 2020
Gedruckt auf umweltfreundlichem, säurefreiem Papier aus nachhaltiger Forstwirtschaft.

ISBN 978-3-7369-7204-9
eISBN 978-3-7369-6204-0

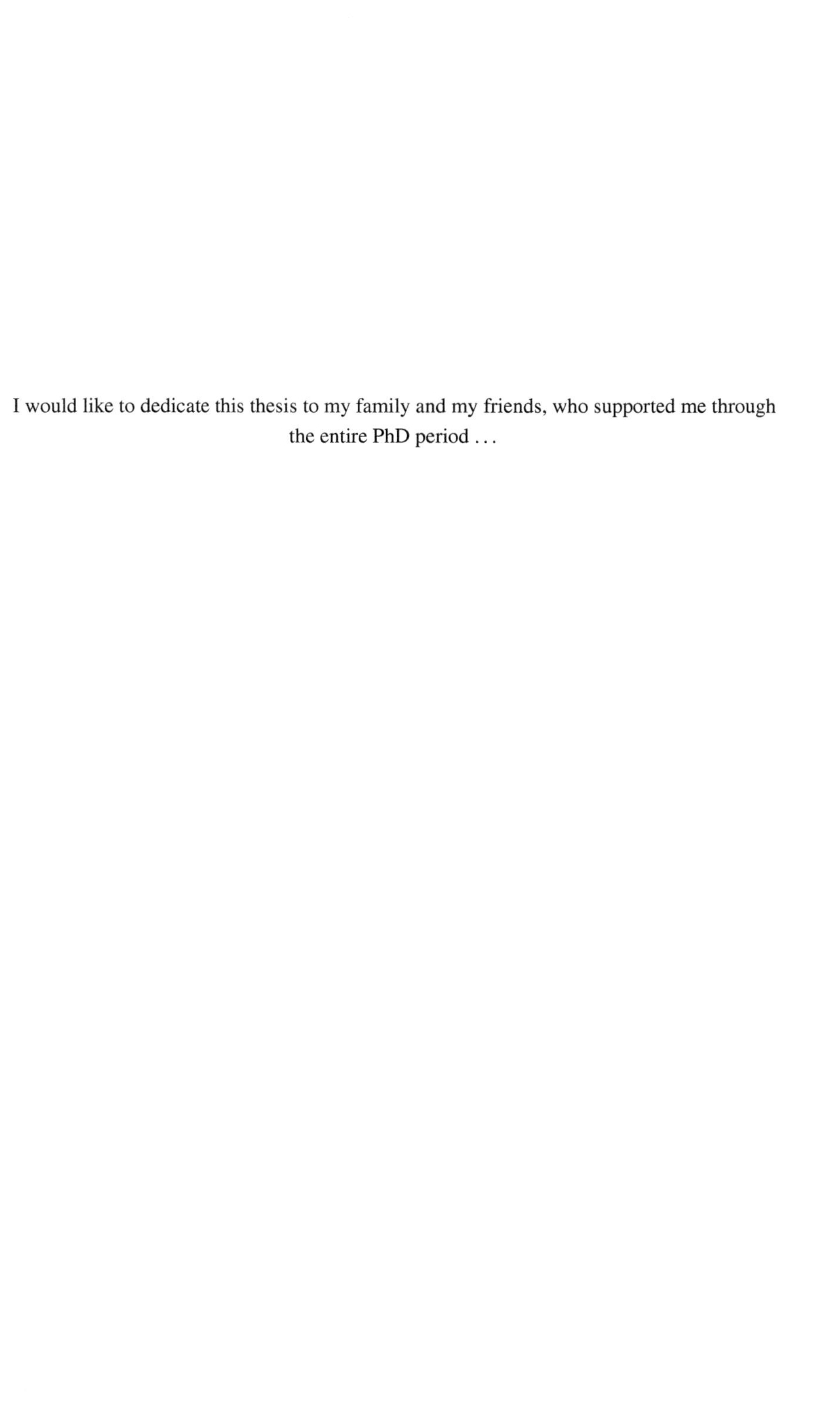

I would like to dedicate this thesis to my family and my friends, who supported me through the entire PhD period . . .

Declaration

I hereby declare that except where specific reference is made to the work of others, the contents of this dissertation are original and have not been submitted in whole or in part for consideration for any other degree or qualification in this, or any other university. This dissertation is my own work and contains nothing which is the outcome of work done in collaboration with others, except as specified in the text and Acknowledgments.

Dimitri Stoppel
Juli 2019

Acknowledgements

I would like to thank all people who help me to get to the point were I am now. First of all, I would like to thank Prof. G. Tränkle, who was also leading the FBH at the time of my stay for supervision my dissertation. The meetings with Prof. G. Tränkle were fruitful and always scientifical and mentally supportive. Many thanks also to Prof. N. Weimann, who leaded the InP Devices and SciFab group. Prof. N. Weimann was for the whole period my first contact for daily advises, furthermore he was an impressive role model in all scientific concerns. Furthermore, I would like the members of my working group, starting with Dr. M.I. Schukfeh, Dr. K. Nosaeva, Dr. S. Monayakul, Dr. S. Boppel, N. Volkmer, M. Brahem and Dr. N. Halder for the great support and the incredible atmosphere. Many thanks to the technicians of the FBH, who where supportive for all requests and problems in my experiments. In particular, I would like to thank S. Hochheim, D. Renter, N. Thiele, K. Kunkel, S. Breuer, A. Kuelberg, J. Behrchen, J. M. Koch, A. Runge, S. Schoenfeld, K. Ickert, M. Matalla, N. Sabelfeld for the daily help and support my work and the great team spirit. I also would like to thank the group leaders of the process technology department for the discussions and consultation, starting with Dr. P. Wolter, Dr. W. John, Dr. L. Weixelbaum, Dr. I. Ostermay, Dr. A. Thies, Dr. R.S. Unger and Dr. O. Krüger. Additionally, I like to thank Prof. W. Heinrich who is leading the microwave department of the FBH and Prof. V. Krozer, who is leading terahertz department for the support. From both groups, I would like to thank the colleagues for the discussions, measurement, simulations and design support, beginning with Dr. M. Hrobak, Dr. M. Hossain, Dr. B. Janke, T. Shivan, A. Raemer and S. Schulz.

Furthermore, I would like to gratefully acknowledge partial financial support under DLR project MIMIRAWE (50 RA 1327).

Finally, thanks to my family, my girlfriend and my friends for mental support, distraction and inspiration during my work.

Zusammenfassung

Anwendungen bei Frequenzen über 300 GHz sind heutzutage noch im Forschungsstadium. Ein Grund für das Fehlen von Anwendungen ist das fehlende Angebot von kommerziell verfügbaren Systemen und Komponenten in dieser Frequenzregion. Ein vielversprechender Kandidat für Hochfrequenzkomponenten ist der Transfersubstrat-Prozess am Ferdinand-Braun-Institut (FBH) in Berlin. Dieser Prozess nutzt die sog. Substrattransfertechnologie (TS), um die monolithisch integrierten Hochfrequenzbauelente (MMICs) auf ein beliebiges Trägersubstrat zu transferieren. Solche Trägersubstrate können zum einen passiv zur mechanischen und thermischen Unterstützung verwendet werden, als auch aktive vollprozessierte Bipolartransitor integrierte Schaltkreise beinhalten. Durch die Verbindung von hochkomplexen Bipolarschaltkreisen mit den hochfrequenten Indiumphosphid-Heterobipolartransistoren (InP HBTs) kann sowohl Komplexität als auch die Leistungsfähigkeit dreidimensional (3D) in einem platzsparenden Prozess integriert werden.
In dieser Dissertation wird der komplette InP TS Prozess Schritt für Schritt beschrieben. Während des Prozessüberblicks werden die drei Teile, Benzocyclobuten (BCB) Trockenätzen, Nickel-Chrom (NiCr) Widerstände und die through-silicon vias (TSV) / Siliziumdurchkontaktierungen aus dieser Arbeit in den InP Prozess eingeordnet.
3D integrierte MMICs, wie z.B. die InP HBTs, sind meist in einem hochfrequenztauglichen Material eingebettet. Das meistverwendete Material zur Planarisierung und zum Verbinden vom InP- und Träger-Substrat ist BCB. In dieser Dissertation wird die Entwicklung eines induktiv gekoppelten Plasmaätzprozesses beschrieben (ICP). Der entwickelte Prozess führte zu einer Verfünffachung der Ätzrate von 60 nm/min auf 300 nm/min bei gleichbleibender Anisotropie und Bias-Spannung. Während der Untersuchungen wurde eine parasitäre Seitenwandabscheidung, bestehend aus AlF_xO_y entdeckt. Die Grundursache für diese Abscheidung konnte auf die Kopplungsplatte, bestehend aus Al_2O_3, zur Induktivantenne des Plasmaätzers zurückgeführt werden. Mittels eines Tetramethylammoniumhydroxid (TMAH) basierenden Entwicklers konnte die Seitenwandabscheidung selektiv entfernt werden, ohne dem MMIC Aufbau zu schaden. Eine Variation der Substratträgermaterialien zeigte einen Einfluss auf die Ätzrate und die Form der Seitenwandabscheidung, wobei eine komplette Unterdrückung nicht erreicht werden konnte. Als Ausblick für kommende Weiterentwicklungen im Bere-

ich des Trockenätzens wird zum Schluss noch eine Methode mit einem Faradaykäfig zur Unterdrückung der parasitären Seitenwandabscheidung motiviert.
Für komplexe Schaltkreistopologien in Hochfrequenzanwendungen sind resistive passive Elemente unabdingbar. In dieser Dissertation wurden drei unterschiedliche Technologiekonzepte zur Herstellung der NiCr Widerstände untersucht. Es wurde ein sehr geringer Kontaktwiderstand mit $8 \times 10^{-10}\ \Omega \cdot cm^2$ zwischen NiCr und dem Kontaktmetall erreicht. Es zeigte sich, dass sich mit der Anbindung des resistiven Materials von unten zum Kontaktmetall und einer Strukturierung mittels lift-off, die besten Ergebnisse erzielen lassen. Die Kontaktierung von oben mit einer nasschemischen Strukturierung (Cr-etch18) des NiCr, erzielt verhältnismäßig hohe Kontaktwiderstände mit $2.61 \times 10^{-7}\ \Omega \cdot cm^2$, die auf eine mögliche Re-Oxidation des NiCr zurückgeführt werden können. Eine massive Unterätzung des NiCr unter einer Siliziumnitrid (SiNx) Schicht konnte bei der Strukturierung des NiCr mittels Nassätzen und einer Bodenkontaktierung beobachtet werden. Dieser Effekt konnte mit dem elektrochemische Potenzial während des Nassätzens zwischen edlen und unedlen Metallen erklärt werden. Thermische Simulationen und Stresstests konnten eine Widerstandstemperatur während des Gleichstrombetriebs von $T \geq 300\,°C$ aufzeigen. Das Technologiekonzept mit den besten Ergbenissen wurde in mehreren MMIC-Prozessen als resistives Bauelement, mit einem Schichtwiderstand von $25\ \Omega/\square$, integriert. Hochfrequenzsimulationen- und Messungen zeigten eine Übereinstimmung der Ergebnisse mit einem Reflexionskoeffizienten von unter 30 dB in einer Frequenzbandbreite von 10 MHz bis 220 GHz. Die Nutzbarkeit der NiCr Widerstände in Schaltkreisen konnte mittels eines Wanderwellenverstärkers (TWA) mit einer Verstärkung von 12 dB bei 95 GHz nachgewiesen werden.
Der Aufbau von Hochfrequenzkomponenten zu einem System erfordert besondere Maßnahmen, um ungewollte parasitäre Moden während des Betriebs zu vermeiden. Im InP Prozess wurden zur Unterdrückung dieser Moden TSVs im Siliziumträgersubstrat integriert. Das Bohren der Löcher, mit einer Tiefe von 200 µm wurde mittels eines ultravioletten (UV) Lasers realisiert. Nach dem Bohren wurden die Löcher mittels eines Hochdruckwasserstrahls mit 20 MPa Druck gereinigt und anschließend mit einem Grundmetall bestehend aus 100 nm Titan und 200 nm Gold versehen. Anschließend wurden die Löcher elektrochemisch mit 10 µm Gold aufgefüllt und mit BCB planarisiert. Die Planarisierung der hoch anspruchsvollen Aspektverhältnisse konnte durch einen vorhergehenden Einweichprozess mittels Mesitylen, erreicht werden. Das überstehende BCB wurde bis zum Silizium mittels Plasmaätzen abgetragen und anschließend mechanisch poliert. Zur Kontaktierung der galvanisierten Seitenwand wurde noch eine Metalldecke mittels lift-off auf die TSVs aufgebracht. Am Ende des InP HBT Prozesses wurde das Substrat auf eine Zieldicke von 700 µm auf 127 µm abgedünnt. Es konnten flächendeckend TSV Durchgangswiderstände von circa 0.3 Ω gemessen werden.

Gleichstrom und Hochfrequenzmessungen auf Waferebende konnten keine negative Beeinflussung der Schaltkreise mit integrierten TSVs nachweisen. Die finalen Substrate wurden vereinzelt und anschließend in einen Radar-Demonstrator integriert.

Abstract

Applications at frequencies above 300 GHz are still at research state due to missing commercially available systems. A promising candidate for the respective semiconductor component is the indium phosphide (InP) transferred-substrate process at the Ferdinand-Braun-Institute (FBH) in Berlin. This particular process utilizes the wafer bonding technique, which allows the transfer of the active monolithic microwave integrated circuits (MMICs) onto a host substrate. Such host substrate can be either a passive substrate that is equipped with through-silicon vias (TSVs) or a BiCMOS wafer. Such hetero-integrated approaches offer ideal conditions to fulfill needs in application with a demand for complexity (BiCMOS) and large bandwidth (InP).

In this thesis, the entire InP MMIC process at the FBH is described step-by-step. Within thesis context, the three topics benzocyclobutene (BCB) dry etch process development, nickel-chrome (NiCr) thin film resistor (TFRs) development and through-silicon via (TSV) implementation are described in greater detail.

3D-integrated MMICs like the InP process at the FBH are often embedded in low-k dielectric materials. BCB is used due to its outstanding thermal, high frequency and mechanical properties. BCB serves as bond material between the InP and the host substrate and is also used as planarization material for the subjacent layers. In this dissertation, a novel etch process with an inductive coupled plasma (ICP) etcher was developed with sulfur hexafluoride and oxygen, with a fivefold etch rate increase from 60 nm/min to 300 nm/min compared to the standard process with a reactive ion etching (RIE) equipment, while anisotropy and substrate bias are maintained. An AlF_xO_y composed etch by-product was found after etching and etch mask stripping. The root cause for redeposition was traced back to the coupling plate of the ICP reactor, which consists of Al_2O_3. With a tetramethylammonium hydroxide (TMAH) based resist developer it was possible to selectively remove the redeposition after the plasma etching. The impact of different carrier materials on the etch rate and by-product formation. A method to suppress the redeposition during the etching was motivated as an outlook by utilizing Faraday shielding.

For circuits, resistive passive elements are inevitable. Three different technology schemes have been evaluated, where the process with a bottom contact to the resistive material and structured by lift-off demonstrated the best electrical results. A NiCr TFR process with lowest contact resistance with $8 \times 10^{-10}\ \Omega \cdot cm^2$ of resistive material to metal was developed. A resistor contact from the top, structured by wet etch (Cr-etch18), indicated a high contact resistance with $2.61 \times 10^{-7}\ \Omega \cdot cm^2$, due to possible re-oxidation of the exposed NiCr or re-sputtered silicon nitride (SiNx). Massive underetched NiCr beneath SiNx was observed during structuring of the NiCr by wet etch, due to the appearing electrochemical potential between the noble handover metal and less noble NiCr. Thermal simulations and stress tests indicated a high thermal resistance, with a temperature of $T \geq 300\,°C$ during operation of the NiCr TFRs integrated on top of BCB. The process scheme with the benchmark results was implemented in several active InP MMIC fabrication runs with a sheet resistance of $25\ \Omega/\square$. Simulations and high frequency measurements are in perfect agreement for $50\ \Omega$ terminations, with an input reflection coefficient below 30 dB in a range from 10 MHz to 220 GHz. The usability in an active circuit was demonstrated in a traveling wave amplifier with a forward gain of 12 dB at 95 GHz.

Mounting high frequency elements into a system can lead to unwanted effects, such as the occurrence of parasitic parallel plate modes. To suppress the unwanted degradation of the performance, an approach with through-silicon vias was followed. Therefore, an ultra-violate (UV) laser was used to drill TSVs with aprox. 200 µm into silicon substrates. After the vias were cleaned with 20 MPa water pressure, they were metalized with a seed metal of 100 nm titanium and 200 nm gold and electroplated with 10 µm gold. Afterwards, a method was developed to planarize the high aspect ratio of the caused topology with BCB. The best results were achieved with a two-step BCB application, first low viscosity BCB followed by prewetting with mesitylene and finalizing with high viscosity BCB. For a suitable flat surface for wafer bonding, the protruding BCB was dry etched with a sulfur hexafluoride and oxygen plasma and mechanically polished afterwards. To short circuit the parallel plate modes, a landing pad on top of the via was applied to establish an electrical connection with the InP MMICs. In a final step, the substrate was thinned down to 127 µm from originally 700 µm. In experiments, a low resistance of circa $0.3\ \Omega$ per via was measured from via top to via bottom. The process was implemented into a full InP HBT process run. High frequency on-wafer measurements indicated no degradation of the device performance with implemented TSVs. The finished substrates were diced and mounted in radar demonstrator modules with transceiver and receiver.

Contents

Chapter 1

Introduction

1.1 Motivation

The world we live in today is dominated by the information technology of the 21st century. The growing demand for more bandwidth in communication enabled the commercialization of monolithic microwave integrated circuits (MMICs). But also other fields now can be accessed with the help of MMICs, e.g. in biology and medicine with the non-invasive blood glucose measurement, protein states, molecular signatures, tissue identification and disease detection [1, 2]. These and many other application have a huge demand for high frequency electronics. To exploit the possibilities for the next generation of electronics in the terahertz frequency (> 300 Ghz), it is necessary to develop and commercialize terahertz microwave integrated circuits (TMICs).
Promising candidates for active devices for TMICs are based on silicon germanium (SiGe), gallium nitride (GaN) and indium phosphide (InP) [3–5]. Groups across the world are researching fast switching devices for the next generation of MMICs.
SiGe heterobipolar transistors (HBTs) are the high frequency (HF) heritage of the classical silicon complementary metal-oxide-semiconductor (CMOS) transistors. In this technology it is possible to integrate both, HBTs and high-electron-mobility transistors (HEMTs) monolithically on one substrate. Most recent results show highest maximum unity current gain cut-off frequency (f_t) and maximum oscillation frequency (f_{max}) at 300 and 500 GHz, respectively. Limited by the bandgap of silicon, the devices are have a comparable low breakdown voltage of around 1.5 V. With this technology scheme it is possible to apply the high integration capability of silicon and combine it with high bandwidth devices, which is an interesting aspect of this approach [5].
Another technology scheme is reaching from the high power area of GaN devices. The latest achievements with GaN HEMTs showing f_t, f_{max} of 400 and 550 GHz. The obvious focus of

Table 1.1 Comparison of state-of-the art mm-wave technologies

f_t [GHz]	f_{max} [GHz]	Breakdown [V]	$L_E/W_E[\mu m^2]$	Technology	Institution	Ref.
337	400	3	0.25x4	InP/GaAsSb HBT	NTT	[7]
360	330	4.5	0.8x6	InP/InGaAs HBT	FBH	[8]
270	750	5	0.25x6	InP HBT	NGAS	[9]
610	1500	3	0.025x10	InP HEMT	NGAS	[10]
428	621	5	0.2x4.4	InP/GaAsSb HBT	ETH	[11]
404	901	4.3	0.18x2.7	InP/InGaAs HBT	UCSB/Teledyne	[12]
400	550	15	0.02	GaN HEMT	HRL	[3]
521	1150	3.5	0.13x2	InP	Teledyne	[4]
300	500	1.5	0.13x2.69	SiGe HBT	IHP	[5]
245	450	1.7	0.15x1	SiGe HBT	IMEC	[13]

GaN transistors are the high power applications, due to the high breakdown voltage and the high power added efficiency (PAE).

Nowadays the fastest oscillation speed achieved with a transistor has been fabricated with the highest electron mobility material, indium gallium arsenide (InGaAs). f_t and f_{max} values of 610 and 1500 GHz have been extracted from measurements of an InP HEMT. With such high bandwidth devices, power amplifier (PAs) with 10 stages were measured at 1.0 THz with a gain of 9 dB [6].

At the Ferdinand-Braun-Institute (FBH) in Berlin, an unique technology approach is followed. Instead of the classical triple mesa transistor scheme with lateral devices, a transferred-substrate (TS) process is utilized to fabricate vertical devices. A vertical approach has the advantage, that it avoids the occurrence of the extrinsic base collector capacitance. Such an approach is possible, since the half finished transistor is transferred to a host substrate to access the collector from the backside. A second and important benefit is the opportunity to either use a substrate with high thermal conductive substrate or e.g. a fully processed bipolar junction transistor with complementary metal–oxide–semiconductor (BiCMOS) wafer [14], as seen in Fig.1.1. Both, the immense scalability of the HBTs and the 3D-integration enable the fabrication of next generation TMICs with high complexity and frequency.

Figure 1.1 Macro photograph of a finished hetero integration wafer (BiCMOS and InP).

For TMICs and MMICs it is necessary to lower the process time to offer affordable circuits with a short design and fabrication loop time. Furthermore it is important to enhance the technology to enable a more complex topology more capabilities.

1.2 Scope of the dissertation

The scope of this dissertation is the improvement and further development of an TMIC process at the FBH. In chapter 1, the introduction for high frequency devices is established as it provides the motivation for this work. Chapter 2 provides a detailed description of the transferred-substrate process at the FBH. The subsequent three chapters show and explain the process developments that have been done to improve the process, constituting the main scientific achievements presented in this dissertation. In chapter 3, the process development for an improved plasma etch recipe is described, which decreases the process time significantly at maintained process quality. The development and implementation of thin film resistors (TFRs) fabricated with nickel-chrome (NiCr), is described in chapter 4. Aspects of process technology, thermal simulation, stress test, high frequency measurements and the usage in circuits are discussed. The third and last part of this thesis is covering the process development and integration of through-silicon vias (TSVs) in the TS process in chapter 5. The necessity thereof is demonstrated using high frequency simulations. Subsequently, the detailed process development and the implementation into the process is covered. The sixth chapter provides a conclusion as well as an outlook into future work.

Chapter 2

InP HBT transferred-substrate process

2.1 Transferred-substrate process

The transferred-substrate (TS) process at the FBH is a unique method to create true vertical InP HBTs. Opposed to the classical triple mesa transistors, the TS process accesses the collector directly from the top due to the fact, that the device is flipped vertically.

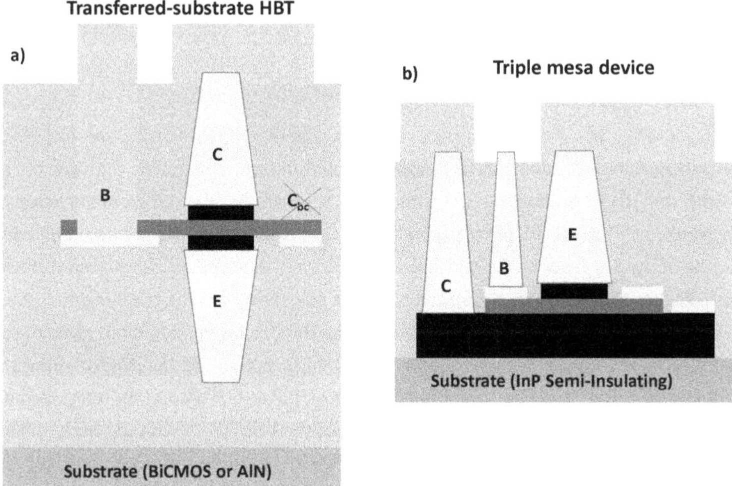

Figure 2.1 Comparison between the triple mesa and the transferred-substrate technology, a) transferred-substrate HBT, b) triple mesa HBT. Based on [14].

The parasitic collector base capacitance C_{bc}, which decreases the unity current-gain cutoff frequency f_t of a transistor, is eliminated in the TS process as seen in Fig.2.1. The extrinsic

C_{bc} is present in the triple mesa approach, while the TS process is released from that parasitic effect.
The common-emitter unity current-gain cutoff frequency f_t is,

$$\frac{1}{2\pi f_t} = \tau_c + \tau_b + C_{bc} \cdot (R_{ex} + R_c) + \frac{\eta kT}{qI_e}(C_{bc} + C_{je}) \tag{2.1}$$

where τ_c and τ_b are defined as collector and base transit times and C_{bc} and C_{je} standing for the depletion capacitances for the collector and emitter. R_{ex} and R_c are the collector and emitter resistances and the $(\eta kT/qI_e)^{-}1$ term is the transconductance of the transistor [15].
The maximum oscillation (unity power-gain) frequency f_{max} depends directly on f_t,

$$f_{max} = \sqrt{\frac{f_t}{8\pi(RC)_{eff}}} \tag{2.2}$$

where RC_{eff} is the base-collector time constant that includes the C_{bc} from Equ 2.1 and the R_{bb} which is the total sum of the base resistances [15]. By assuming that the base resistance is much larger than the emitter and collector resistances, their effect becomes negligible and only the base-collector path contributes to the $(RC)_{eff}$. With this assumption the f_{max} can be described as,

$$f_{max} = \sqrt{\frac{f_t}{8\pi(R_{bb}C_{bc})_{eff}}} \tag{2.3}$$

where R_{bb} and C_{bc} has the most impact on the f_{max} of the HBT. With the minimized C_{bc} in the TS process, the contribution to f_t and f_{max} is minimal. This paves the way for higher bandwidth InP HBT compared to the triple mesa transistor, at the same emitter width.
The minimized C_{bc} in the TS process is possible, due to the full wafer bonding procedure. In the middle of the TS process, the half finished HBT is transferred by adhesive wafer bonding to a host wafer. This wafer transfer facilitates the possibility to attach a non-processed host wafer like silicon (Si) or aluminum nitride (AlN) to the InP wafer. The more groundbreaking opportunity is to replace the host wafer with a fully processed BiCMOS wafer [8, 14]. With these technology schemes combined, it is possible to use the highly integrated silicon technology, so to say the brain with the muscles of the InP HBTs with the high breakdown voltage and the high frequency capability. With those technologies combined, it becomes possible to integrate not laterally but vertically (3D) e.g. voltage-controlled oscillator (VCO) with amplifier chains [14].

2.2 Process description

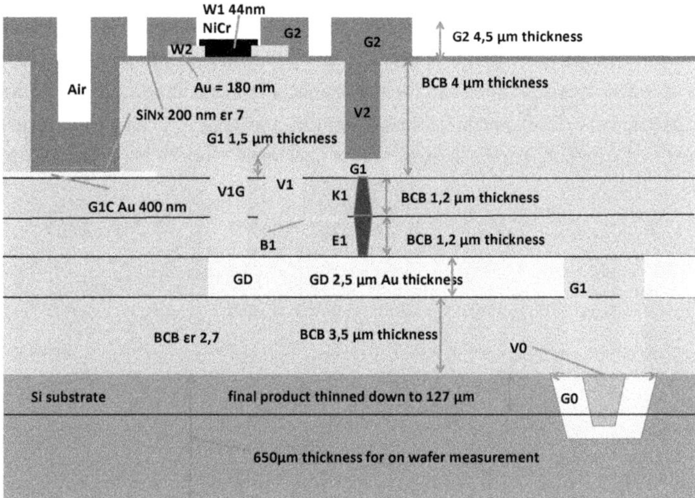

Figure 2.2 Cross section view of the TS process with a silicon host substrate with TSVs, which can be substituted by a BiCMOS wafer. Based on [16].

In the present status, the TS process contains three layers of electroplated gold metals (GD, G1 and G2) as shown in the Fig.2.2. These metals form the ground metal and connection to the emitter (GD), the first interconnects to emitter, base, collector, host substrate and capacitors (G1). The microstrip line, measurement and bond pads and the interconnects to the previous layers are connected with the last electroplated metal (G2). The process contains a metal-insulator-metal (MIM) capacitor with silicon nitride (SiNx) as dielectric material with a permittivity of $\varepsilon_r = 7$. A NiCr thin film resistor with a sheet resistance of 25 $\Omega/\square$ is integrated at the top layer. It is possible to integrate through-silicon vias into the host substrate to suppress parasitic substrate modes, that can occur after dicing the MMICs and mounting them into a system. The whole device and all its passive components and interconnects are embedded in benzocyclobutene (BCB). BCB has very good planarization capabilities, a high glass transition temperature of about 350 °C and with a ε_r of 2.65 it is a suitable material for high frequency applications [17].

The process begins with creating a mask layout. In the design phase the circuit designer utilizes either the extracted models of the active and passive components from a previous wafer run or an estimated model if e.g. a new epitaxy is tested. After combining all components, the layout has to be exported. The finished mask layout is created according to the design rules and process bias based on the technology. An example for a final layout with more then 20 lithographic layers is shown in Fig.2.3. This layout contains all necessary components for a successful MMIC process, like process control monitoring (PCM) structures for metal contacts, sheet resistances and isolation measurements. The layout also includes all technology structures e.g. for step-height, alignment check and critical dimension (CD) measurements. To characterize the HF devices and passives, calibration structures are positioned in the layout. The most mask area is consumed by the end products (MMICs).

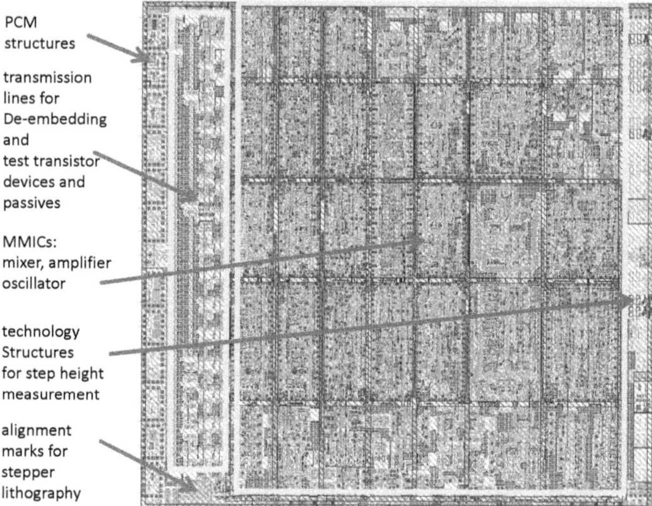

Figure 2.3 Layout example of an TS process with remarks on the content.

The transferred-substrate technology starts with the growth of the epitaxy on InP wafers. In the case of FBH, the epitaxy is ordered from a commercial supplier with following layers from bottom (substrate) to the top: two etch stops, sub-collector, collector, delta doping, grade, setback, base, emitter and emitter cap [18].

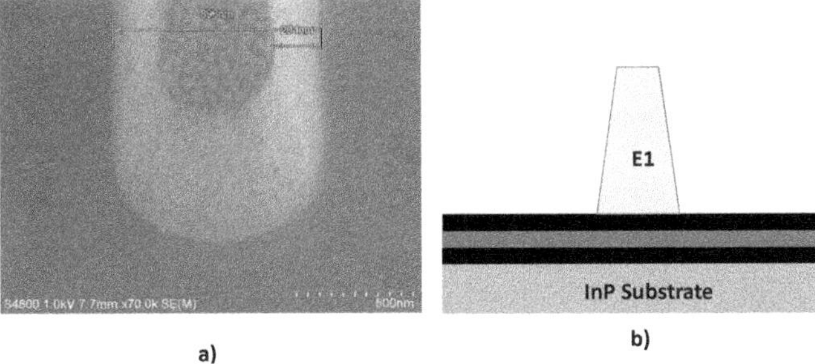

Figure 2.4 Illustration of the emitter process, a) SEM micrograph of a metalized emitter, b) cross-sectional sketch of the emitter.

The process starts similar to the conventional triple mesa process. As first lithography layer, the emitter contact metal is defined by the lift-off technique (E1). The emitter metal sets an ohmic contact to the highly doped emitter cap with extra metal to increase the distance between the forthcoming ground metal and the base as seen in Fig.2.4. The applied metal also serves as an etch mask to structure the emitter semiconductor. With that mask, the emitter cap (InGaAs) is wet etched by sulfuric acid and emitter (InP) by hydrochloric acid. Both acids are highly selective etchants to the target material. With the structured emitter semiconductor, the base contact surface is revealed and the isotropic etch mechanism creates a necessary undercut of circa 50 nm for the following step.

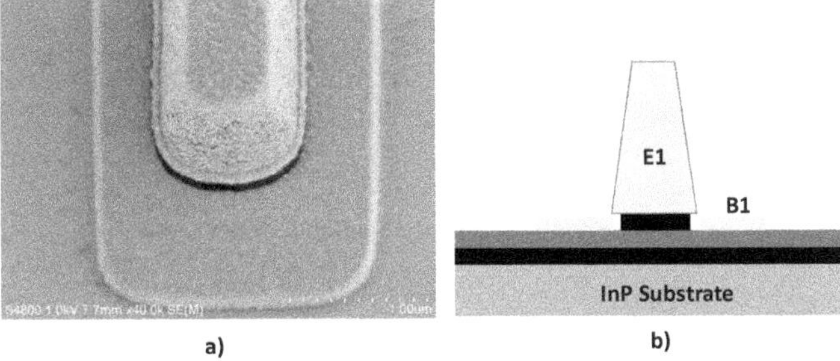

Figure 2.5 Illustration of the self-aligned base process, a) SEM micrograph of a metalized base, b) cross-sectional sketch of the base.

Subsequently, the base metalization (B1) is performed in a self-aligned process as seen in Fig.2.5, in which the base contact is established. This works because of the undercut of the emitter and the thinner base metal compared to the emitter semiconductor thickness. The pretreatment before the metalization and the metal sheet resistance have a significant impact on the f_{max}.

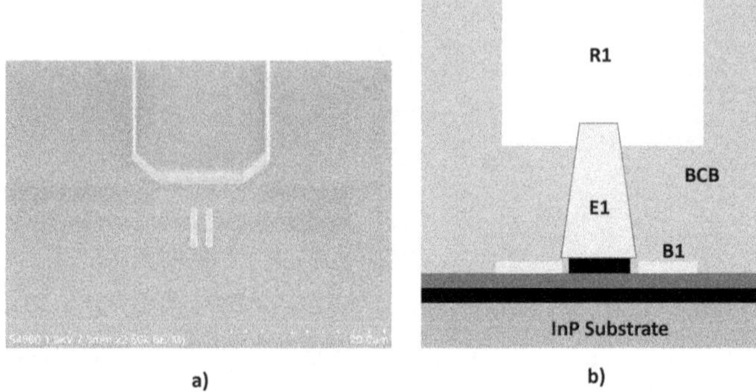

Figure 2.6 Illustration of the planarization and emitter reveal process, a) SEM micrograph of the dry etched BCB, b) cross-sectional sketch of the planarization.

After the base contact is established, the exposed semiconductor of the emitter and base is encapsulated with SiNx, to prevent the emitter-base diode from oxidizing. With the finished diode, the first planarization with BCB follows. To reveal the emitter top and provide a suitable trench for the upcoming ground metal, the BCB is dry etched (R1) as seen in Fig.2.6. The detailed description and development for a suitable dry etch recipe is described in chapter 3. In this step it is necessary to have a sufficient height of the emitter sticking out of the BCB for a good emitter-ground metal contact. Therefore, the etching must be maintained within a sufficient depth but at the same time shallow enough, to avoid a short circuit to the base.

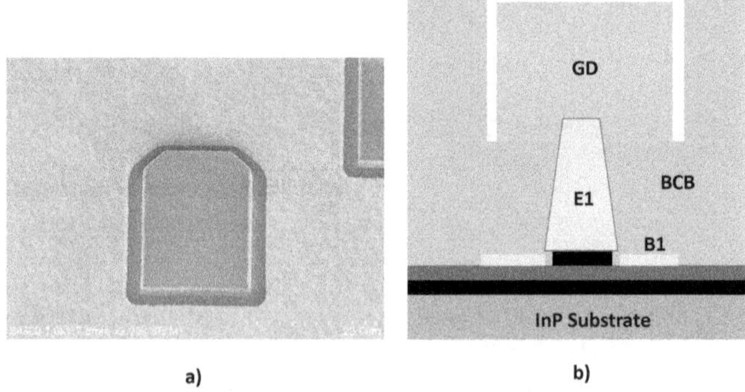

Figure 2.7 Illustration of the GD process, a) SEM micrograph of the electroplated ground metal, b) cross-sectional sketch of the grund metal.

With the emitter heads sticking out of the BCB and a dry etched trench, the ground metal (GD) can be applied by electroplating. The ground metal needs to be on the same vertical level as the roof of the BCB, to provide a flat surface for the following wafer bond process. The GD metal serves for several needs like emitter connection periphery, ground shielding at the high frequency usage and as a heat sink for the HBTs as depicted in Fig.2.7.

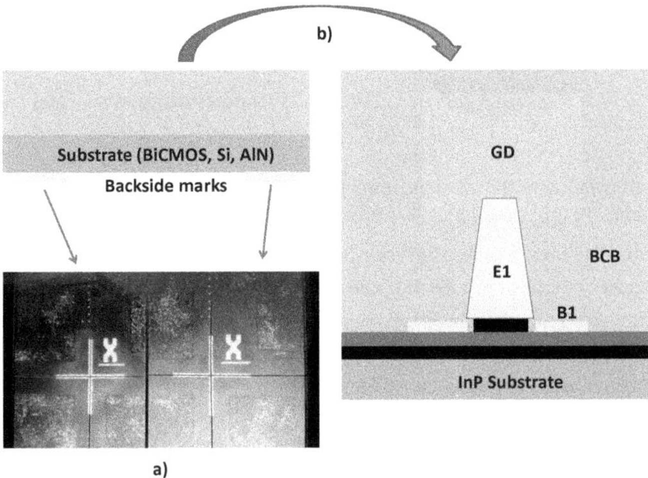

Figure 2.8 Illustration of the wafer bond process, a) photograph of the bond alignment between two wafers, b) cross-sectional sketch of the wafer bond process.

With a uniform and homogenous GD growth, the process is ready for the last steps before wafer bonding. A last planarization with BCB before wafer bonding is applied, to level the small offset between BCB and GD roof. The necessary work to prepare the surface for a host substrate with TSVs, is described in chapter 5. In the case of the BiCMOS host substrate, a barrier metal needs to be applied on the aluminum pads to avoid gold diffusion. In the next step, the bond BCB is applied on both, the InP wafer and the host substrate. Pretreatments are necessary to accomplish a bubble free wafer bond [19]. Before the bond process itself, the wafers have to be aligned to ensure that the hetero integration vias are connecting the host substrate with the InP wafer, as seen in Fig.2.8. The BCB to BCB bond is created by applying temperature and force on the substrate pair.

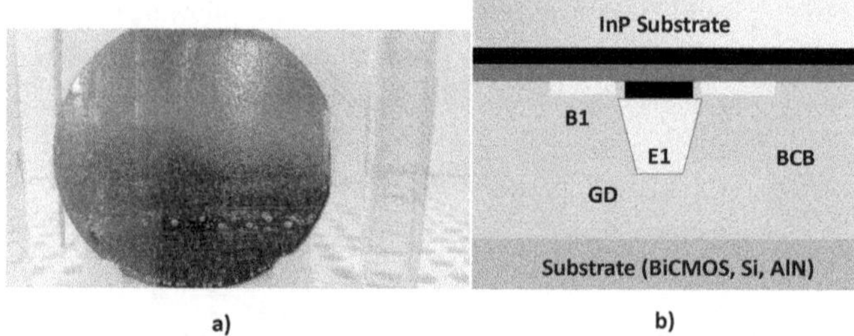

Figure 2.9 Illustration of the substrate removal, a) photograph of a wafer in hydrochloric acid for InP etching, b) cross-sectional sketch of the substrate removal.

Subsequently, the now obsolete InP substrate is removed with heated hydrochloric acid, which selectively etches only the InP and stops on the InGaAs etch stop, as depicted in Fig.2.9. After removing the whole InP substrate, the front side alignment marks have to be opened for an accurate alignment in the following collector process. Furthermore, the etch stop has to be removed to access the collector cap for an ohmic contact.

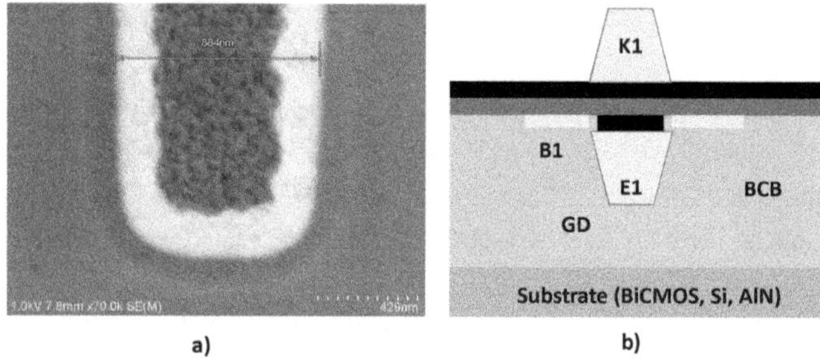

Figure 2.10 Illustration of the collector process, a) SEM micrograph of a metalized collector, b) cross-sectional sketch of the collector process.

With opened marks, the collector is directly aligned on top of the emitter and base. The metalization and mesa etch process is the same as in the emitter process. Here, the metal contact (K1) is established by utilization of the lift-off technique as seen in Fig.2.10. In the following mesa etching the subcollector, collector and the deltadoping is selectively wet etched with hydrochloric and sulfuric acid.

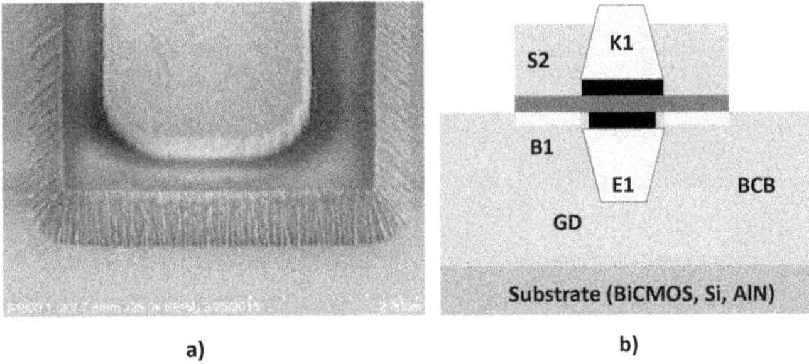

Figure 2.11 Illustration of the base structuring process, a) SEM micrograph of the dry etch base, b) cross-sectional sketch of the base structuring process.

After structuring, the collector needs to be encapsulated. This is done by transforming the pattern into BCB which also serves as a mask (S2) for base dry etching as depicted in Fig.2.11. With that step, the complete HBT device is fabricated.

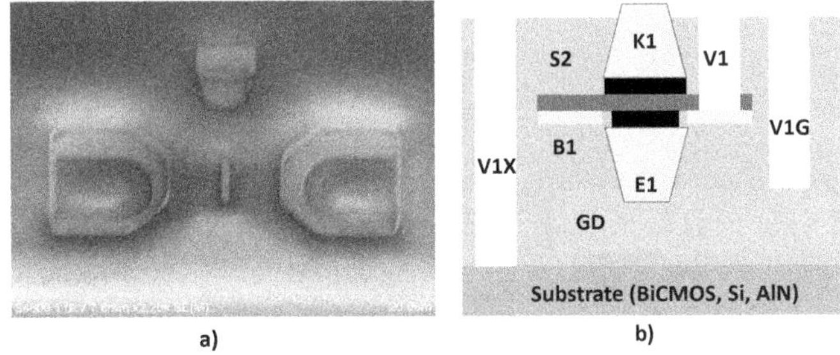

Figure 2.12 Illustration of the via etching processes, a) SEM micrograph of dry etched vias to base and ground metal, b) cross-sectional sketch of the via etching.

In the next step, the transistors again are planarized with BCB. Now the connections to all previous layers are established, as seen in Fig 2.12. Here the interconnects to the host substrate (V1X), to the ground metal (V1G), the base metal (V1) and the top of the collector metal head are established.

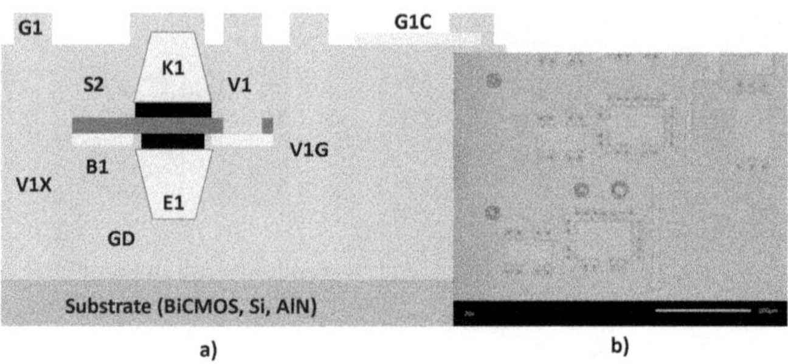

Figure 2.13 Illustration of the second electroplated metal, a) cross-sectional sketch of the second electroplating process, b) micrograph after electroplating.

After the dry etched vias, the bottom metal pad of the MIM capacitor (G1C) is structured by lift-off as depicted in Fig.2.13. Furthermore, the first interconnection metal (G1) is electroplated to connect the host substrate, the GD metal, the base, the collector and the bottom MIM capacitor pads.

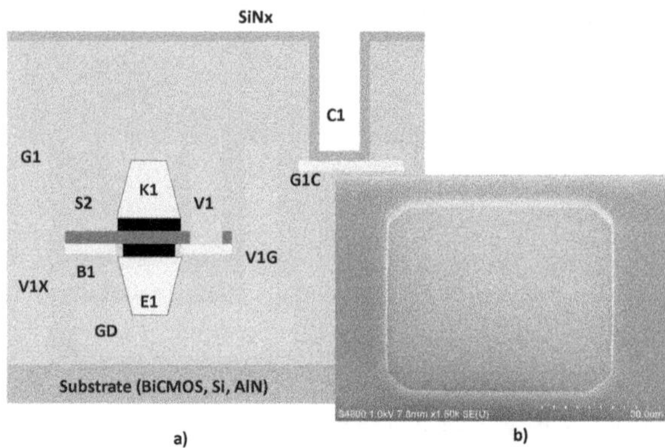

Figure 2.14 Illustration of the MIM capacitor area structuring, a) cross-sectional sketch of the last planarization and MIM capacitor dry etching, b) SEM micrograph after structuring the capacitor area.

Subsequently follows the last planarization step in the MMIC process with BCB. Here additional distance is added to reduce the parasitic capacitance from the microstrip line to GD and the device itself. After the planarization, the capacitor area is defined by dry etching the BCB to the bottom pad of the MIM capacitor (G1C). Afterwards the MIM dielectric (SiNx) is deposited by plasma-enhanced chemical vapor deposition (PECVD), as seen in Fig.2.14.

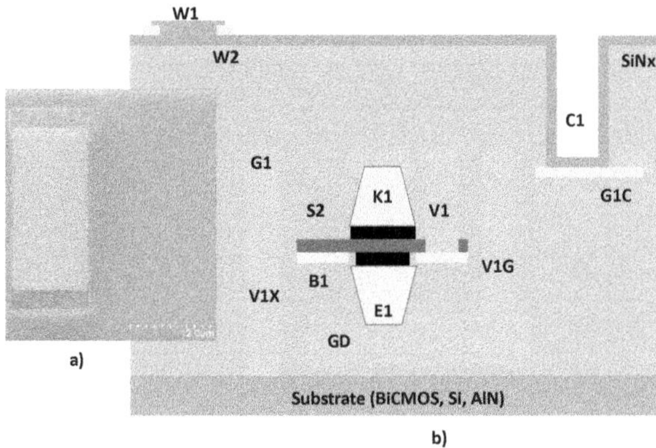

Figure 2.15 Illustration of the NiCr TFR fabrication, a) SEM micrograph of a NiCr resistor with it's contact pads, b) cross-sectional sketch of the TFR process.

On top of the SiNx, the TFRs made of NiCr are applied on the wafers, as depicted in Fig.2.15. Therefore contact pads (W2) and the NiCr (W1) are deposited, respectively. The detailed description of the process development and implementation into the process is discussed in chapter 4.

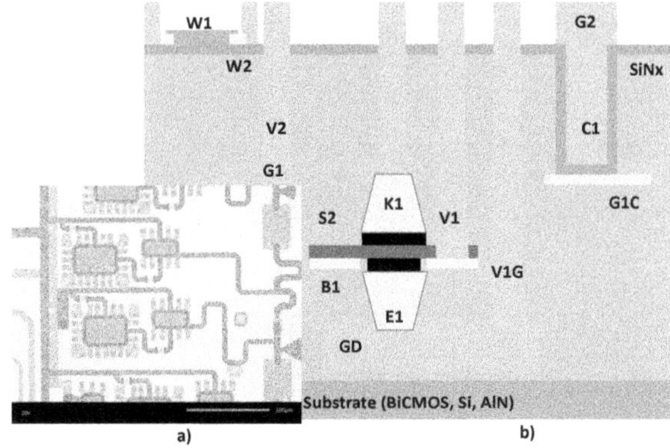

Figure 2.16 Illustration of the last electroplating process, a) micrograph after electroplating an MMIC, b) cross-sectional sketch of the last interconnection metal.

Finally, the interconnections and microstrip lines are fabricated. Therefore the last vias (V2) are dry etched to connect the precedent metal layer (G1). Afterwords the last electroplated metal (G2) is grown on the wafer to connect the TFRs, the MIM capacitors and the previous layer, as seen in Fig.2.16.

At this point, the wafers are ready for on-wafer PCM and HF measurements. For integration into a system, dicing streets (SB) have to be etched into the BCB and in the case of flip-chip mounting also metal bumps have to be applied [20].

Chapter 3

Plasma etching of benzocyclobutene

3.1 Dry etched interconnects in the InP transferred-substrate process

In chapter 2 it is pointed out, that the TS process needs numerous encapsulations with BCB for the fabrication of the MMICs. As seen in Fig.3.1, the TS process beholds a lot of etch steps for interconnects and planarization steps. Following process steps are used in the process, with etch depth of the BCB: R1 (2.8 μm), S2 (1.7 μm), V1 (1.2 μm), V1G (1.2 μm), V1X (7.6 μm), R2 (1.4 μm), C (3.5 μm), V2 (3.0 μm), SB (13 μm) with a total sum of more than 35 μm dry etched depth for each wafer.

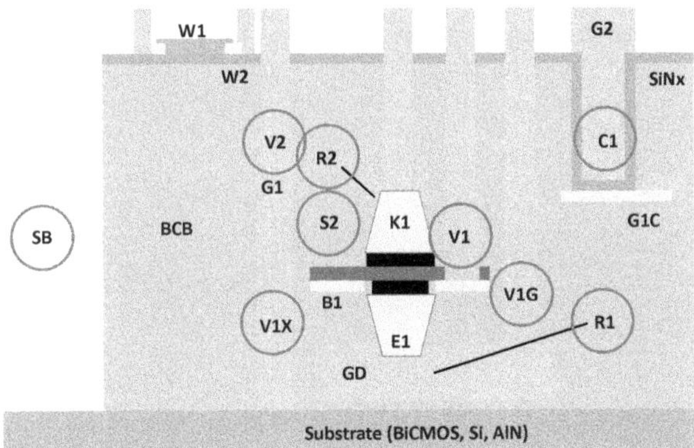

Figure 3.1 Illustration of the significant etch steps in the TS process.

Numerous InP HBT integration schemes rely on BCB as a dielectric layer and for planarization [21, 12, 22, 9, 4]. In the TS process, the BCB is additionally used for adhesive wafer bonding. BCB, also known as benzocyclobutene (BCB) is a low-k polymer, which is perfectly suited for THz applications because of its low loss tangent of 7×10^{-3} at 1 THz [23]. Additionally, BCB has a very low leakage current of 6.8×10^{-10} A/cm^2, high breakdown field of 5.3×10^6 V/cm and almost no moisture absorption of < 0.2 %. Its low processing temperature (around 250 °C) enables BCB as a perfect candidate for THz applications. Due to the high melting point of 350 °C, BCB can be used within a wide process temperature window [17].

Due to the high demand for BCB dry etch steps, a reliable and fast etch process is needed. Ideally, this process should be able to run on several etch reactors to increase process availability by creating system redundancy.

3.2 Plasma for semiconductor processing

3.2.1 Plasma generation and basics

Plasma is the fourth state of matter. Unlike the other states like solid, liquid and gas, the plasma does not occur under normal circumstances on earth. To generate a plasma, a high temperature can be used, like on the sun. There, the atoms are split into free negative charged electrons and positive charged ions [24].

Plasma has been used in the semiconductors industry since the 1960s when it was initially used to remove organic photoresist. Nowadays plasma processes are indispensable in semiconductor technology to structure materials, where classical wet etch solution cannot satisfy the specific requirements anymore. [24].

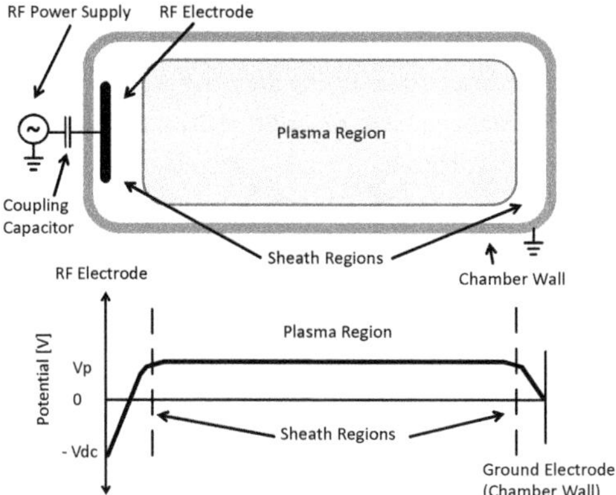

Figure 3.2 Illustration of the dry etch process in a plasma reactor. Based on [24].

To utilize plasma for semiconductor processing, a glow discharge plasma is used [24]. The obvious benefit of using a plasma is the high-temperature-type chemical reactions while maintaining a low substrate temperature. With such ignited plasma in a reactor, reactive species from the feed gas are used to etch the target material. A simple way to generate a plasma, is to apply a radio frequency (RF) field on a ionizable gas in moderate vacuum of around 1 mbar in a plasma reactor. A simple configuration of a plasma reactor is depicted in Fig.3.2.

In the phase of plasma ignition, the RF field applies a force on the electrons inside the atoms/molecules, which are moving faster than the bigger positively charged ions. This leads to the phenomenon, that the electrons can follow to RF field and collide with the slower positive ions (elastic collision) on their way. Four cases are possible when electrons are interacting with atoms/molecules [25]:

- Case 1: electron energy is too low for dissociation and ionization (elastic collision)
- Case 2: electron energy is too low for dissociation and ionization, but enough for stimulation (light emission)
- Case 3: electron energy is too low for ionization, but enough for dissociation (radical)
- Case 4: electron energy enough for ionization (ions)

With this bombardment on the molecules, the plasma is heating up until an equilibrium sets between the ionization and recombination of the gas. The coulomb forces between the electrons and positive ions create a field that ensures a self-limiting process, that stops when the RF power is off. The substrate is usually placed on the RF electrode. The walls are grounded, to act as a grounded electrode. Inside the plasma region, the molecules maintain charge neutrality, for a consistent equal number of positive and negative ions. To ensure a fully functional and continuously burning plasma, a sufficient and consistent RF power (Vp) has to be applied on the RF electrode [25].

Outside the plasma region exists the sheath region. All potential drops occur in the sheath region. On one hand, the electrons leave the plasma region and move to the reactor wall, charging the wall negatively. This negatively charged wall reflects further electrons, creating a self isolating barrier to prevent the plasma from discharging. On the other hand, the positively charged ions are attracted and are moving towards the negative wall. This sheath thickness depends on many factors, like electron temperature, RF power, density and many more.

A negative potential (-Vdc) on the RF electrode is a self-induced direct current (DC) bias. This bias is necessary to balance the faster electrons (smaller mass) flux against the slow ion flux, to achieve a time-average neutrality at the electrode. In the processing, the DC bias is an indicator for the energy of the incoming ions to the electrode. Whereby, the DC bias depends on the power, pressure, gas, electrode geometry and the electrode surface. As depicted in Fig.3.2, a coupling capacitor is implemented in the matching network, which acts as a conductor for the RF and simultaneously as an insulator for DC. Consequently, the self-induced DC bias persists besides the RF field. The significantly lower potential of the RF electrode, compared to the reactor wall, leads to an attraction of the positive charged ions towards the electrode, where the etchable substrate is located [24].

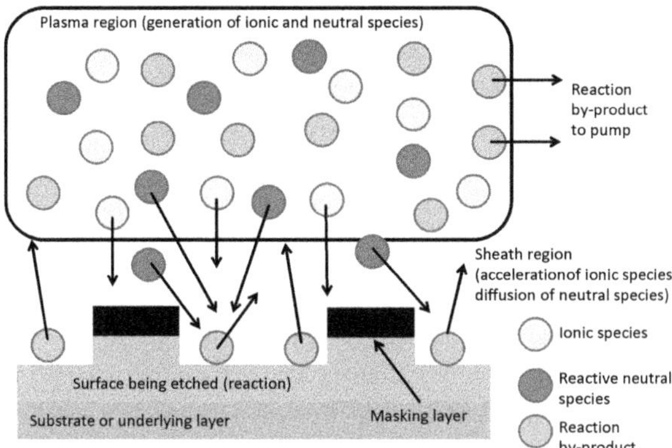

Figure 3.3 Illustration of the dry etch process in a plasma reactor. Based on [24].

The plasma in the the reactor is only partially ionized. Only a percentage part of the molecules is ionized in the plasma region. As depicted in Fig.3.3, the plasma can be divided into three species, except the not ionized and not contributing gases. The species consists of ionic species (ions), namely reactive neutral species (radicals) and the reaction by-product. Ionic species are positive charged ions which are accelerated towards the RF electrode by the applied negative potential (-Vdc as shown in Fig.3.2). The radicals are not considerably charged and can be assumed as neutral molecules. Those radicals are highly reactive and are moving only due to the diffusion mechanism. The third species is the reaction by-product, which has to be volatile, therefore ideally a gas product that can be pumped out of the reactor. The etch mechanism can be described in five steps: (1) generation of radicals and ions, (2) diffusion of radicals and acceleration of ions towards the RF electrode, (3) reaction (etching) of the reactive species with the target material, (4) desorption of the by-product from the surface into the reactor, (5) pumping out of the by-product from the reactor [24]. As indicated in Fig.3.3, the neutral radicals can diffuse at any angle to the surface, because they can not be affected with electric or magnetic fields. Even though the ions can also diffuse to the target material, the main mechanism of the charged ions, is the directional acceleration by the electric fields towards the RF electrode. The by-products are removed only by diffusion mechanism. Many processes in a plasma reactor are still not well understood, which leads to empirical process development in most cases [24].

3.2.2 General parameters for semiconductor plasma processing

When talking about plasma etching, numerous parameters are important for the processing: anisotropy, etch rate, selectivity, uniformity, particles, plasma damage and microloading. The plasma processes are complex and not completely understood, but some general guidelines are available.

Anisotropy (A) describes the etch profile geometry in its cross section

$$A = 1 - l/h \tag{3.1}$$

where l is the lateral etched length underneath the mask and h is the vertical etch depth into the target material [24]. The anisotropy depends on multiple factors in an etch reactor. The substrate temperature is one factor, which also has a huge impact in wet etching. With high substrate temperature, the chemical etching (isotropic profile) is stronger than the ionic etching (anisotropic profile), as seen in Fig.3.4.

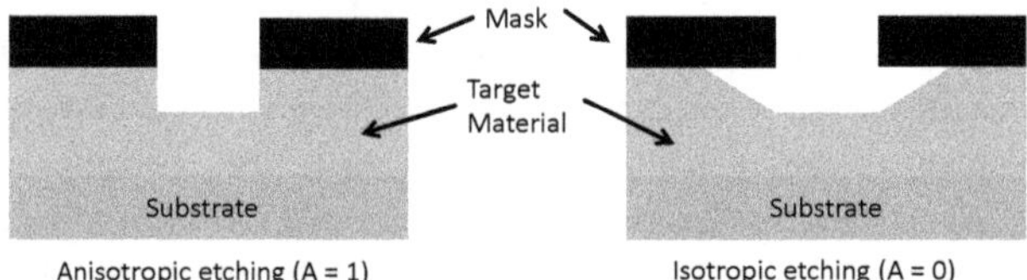

Figure 3.4 Illustration of the dry etch process with an an- and isotropic etch profile

In a capacitive dry etch system it can be assumed, that with higher RF power, simultaneously the anisotropy increases. This is due to the fact, that with more RF power also more electrons are generated and because of this, also the induced self-bias increases, which leads to higher acceleration towards the electrode. At lower pressure, the anisotropy also goes up, because less radicals can diffuse through the sheath region (mean free path). Depending on the target application, either high or low anisotropy is intended [25].

The etch rate also depends on substrate temperature, RF power and pressure. An increase in the values of these parameters leads to an increased etch rate. With rising pressure, the etch rate increases until a diffusion (reaction) limit is reached. Often, the gas mixture in a plasma reactor consists of a few different gases. This is necessary to control the etch rate, by diluting the reactive gas in a non reactive or less reactive gas.

The selectivity (S) is the ratio between two etch rates

$$S = r_{target}/r_{mask} \tag{3.2}$$

where r_{target} is the etch rate of the target material and r_{mask} the etch rate of the mask. A high selectivity is desirable, to avoid the risk of consuming the whole mask and subsequent etching of the target material that was intended to be covered by the mask [24]. The selectivity, the etch rate and the anisotropy depend on the RF power, substrate temperature and additionally on the etch chemistry. Therefore, a chemically stable material is selected for the mask to avoid chemical etching thereof, leaving only physical erosion from colliding particles, which will not react with the etch gases but only gets physically eroded by the accelerated ions.
The uniformity of an etch process usually worsens with increasing wafer sizes [24]. However, the uniformity can be controlled with the reactor design. Here for instance, the electric and magnetic fields can have an influence on the uniformity. Furthermore, the gas inlets and the pump exhaust are important factors for an uniform etch process. This is due to the fact, that the diffusion mechanism of the etch species and the etch by-product through the sheath region, is mandatory for a continuously uniformal etch process.

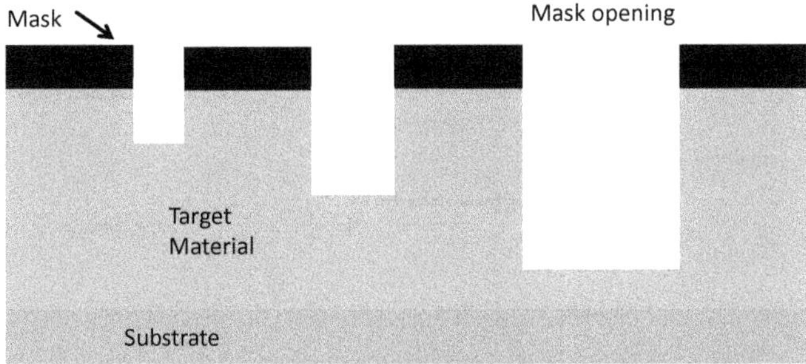

Figure 3.5 Illustration of the microloading effect during plasma etching

Microloading effect, also known as aspect ratio dependent effect (ARDE), describes a phenomenon, which leads in a plasma process to a dependency between etch depth and mask opening [26]. A result of the microloading effect is illustrated in Fig.3.5. This effect can be explained with the diffusion of the reactive species through the sheath region. The lager openings allowing the atoms and molecules with the comparably unobstructed geometry to diffuse easily towards the target material than the smaller openings. This effect leads to the necessity of over etching (etch more than necessary) when processing dry etched interconnects.
Particles in an etch reactor are of particular interest, especially in the high-density production. Particles can result from composition of the reaction gases and the by-products. Therefore, it is often mandatory to perform a cleaning plasma after each etch run [24]. Not only particles, but also thick films can occur in such etch reactors. For this reason, many manufacturers are

implementing liners inside the reactor, which protect the chamber wall from contamination. Those liners can be easily exchanged without risk of long downtime.
Plasma damage especially while structuring active devices are of major concern. When high energy ions are hitting the substrate surface, degradation and contamination of the semiconductor can occur [27]. Often a thin protective mask, for example a SiNx layer, prevents the device from damage. Additional the ultra violet (UV) radiation can damage the semiconductor and positive photo resist masks [24].

3.3 Capacitive and inductive coupled plasma etch equipment

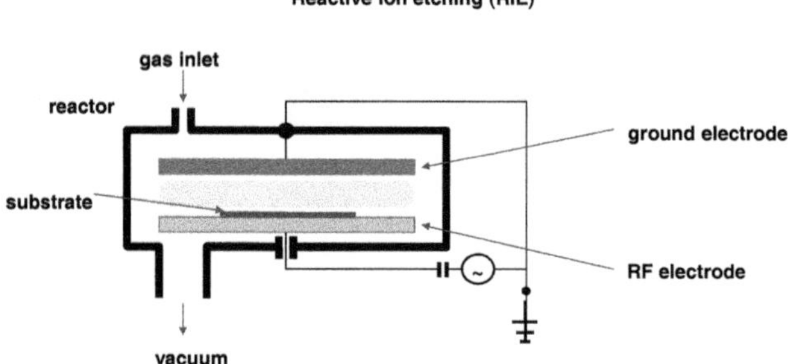

Figure 3.6 Principal RIE schematic construction. Based on [25].

For the following part of this chapter it is mandatory to understand the principle behind the capacitive coupled plasma etching, also known as reactive ion etching (RIE) and the inductively coupled plasma etching (ICP). As depicted in Fig.3.6, the RIE etcher is similar to the parallel plate plasma etcher. It has a capacitively coupled electrode to avoid DC discharges. This is the classical approach to dry etch semiconductors. When using this equipment, the most important parameter is the RF power, where the ion energy and plasma density is connected. Other parameters like pressure, gas composition and geometry are also important to adjust the anisotropy and selectivity. In general it works similar as the basic plasma reactor described in Chap. 3.2.1. This schematic overview of Fig.3.6 reflects the SI 591, from Sentech Instruments GmbH, which was used in the following RIE experiments.

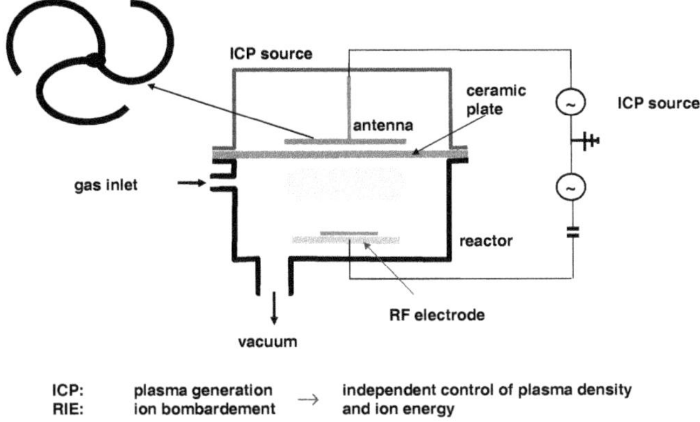

Figure 3.7 Principal ICP schematic construction. Based on [25].

A more advanced plasma etcher is the ICP reactor shown in Fig.3.7. In principle it can be seen as an RIE reactor with an inductively coupled attachment on top of it. With this configuration it is possible to adjust the power of the RF electrode and the ICP source, which enables the opportunity to control the ion energy (bias) and the plasma density (amount of free radicals) separately. The spiral antenna that is shown in Fig.3.7 applies a magnetic field on the plasma, which forces the electrons to follow on a much longer path thereby hitting more atoms. This increased bombardment results in a significantly higher etch rate at lower bias (less plasma damage). Often the ICP antennas are twisted around the ICP hat, but in this configuration the antenna is isolated by a ceramic plate (quartz or AlN) from the plasma reactor to avoid DC discharges. This schematic overview of Fig.3.7 reflects the SI 500, from Sentech Instruments GmbH, which was used in the following ICP experiments.

3.4 Process development

3.4.1 Methods and Tools

Additional to the plasma etcher mentioned in Chap. 3.3 further equipment was used for the dry etch process development:

- reflectometer (RM2000, Sentech Instruments GmbH)
- ellipsometer (SE400adv, Sentech Instruments GmbH)
- scanning electron microscopy (Hitachi S4800)

- in-situ process control was performed with a laser interferometer (NANOMES, GFMesstechnik GmbH)
- i-line step-and-repeat lithography tool (Nikon i12)
- dedicated BCB curing oven (YES-PB6-2PCP)
- scanning electron microscopy with an energy-dispersive X-ray spectroscopy sensor (EDX) (Zeiss Ultra+)

Samples for the experiments were prepared on whole 3 inch < 100 > semi insulating silicon (Si) wafers. On those samples a layer stack was created to mimic the actual via process on InP substrates. The experimental setup is depicted in Fig.3.8.

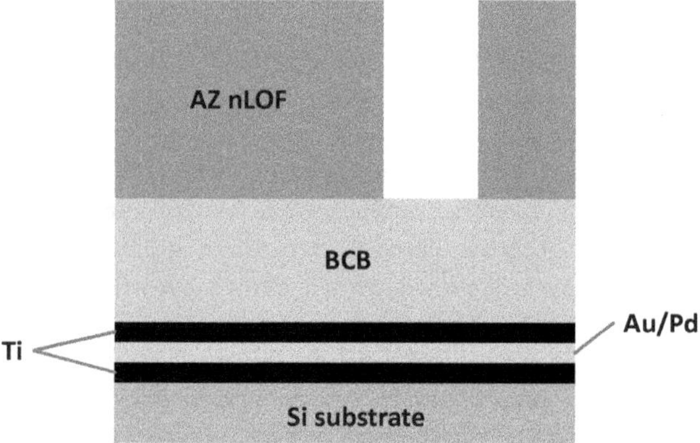

Figure 3.8 Cartoon of the experimental sample setup. Based on [28].

After cleaning the silicon wafers with ultra pure water rinse and oven drying, the metal stack was deposited on the front side (one side polished wafers) using physical vapor deposition. The metal stack consisted of (bottom to top) titanium with 30 nm for adhesion, 80 nm of gold (in chap. 3.4.3) or 40 nm of palladium (in chap. 3.4.4) as an effective etch-stop and an additional adhesion metal for BCB with 20 nm Ti. On top of the metal stack two variations of BCB were deposited (CYCLOTENE 3022-35 and -46) with 2.5 μm and 4 μm thickness for planarization and via etching, respectively. The BCB was cured in nitrogen (N2) atmosphere for 90 min at 240 °C. For the planarization experiments (etch homogeneity across the whole wafer), no mask was applied on the wafer. The AZ nLOF 2035 (distributed by MicroChemicals GmbH) with a thickness of 5.5 μm was used as an etch mask for the via experiments.

3.4.2 BCB parameters and basic dry etch principles

BCB is a polymer that has superior thermal properties besides other reasons, due to the implied silicon inside the cross linked 3D network. The principle structure is shown in Fig.3.9. Benzocyclobutene consists of a cyclobutane ring alongside with single and double bond with the silicon (Si) and methyl group (CH_3) cluster in a cross linked 3D network. The contained Si inside the network enables high process temperatures after cross linking, but also accompany the necessity for fluorine gas to dry etch it.

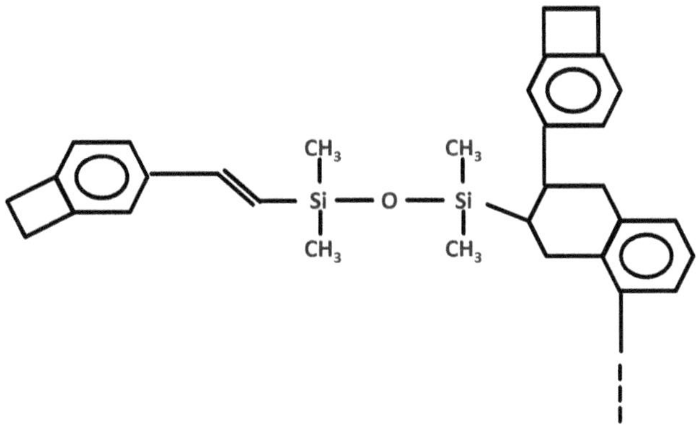

Figure 3.9 Chemical 3D network of cured BCB. Based on [29].

To etch (transform the cross linked network to volatile gases) only two reactive elements, namely fluorine (F) and oxygen (O) are needed. Since the BCB consists of three different elements, namely carbon (C), hydrogen (H) and silicon (Si). Suggested gases to dry etch BCB by the manufacturer are fluorine radical generating gases such as SF_6 , NF_3, C_4F_8, CF_4, and C_2F_6, mixed with O_2 [17]. A possible way to etch BCB would be with sulfur hexafluoride (SF_6) mixed with oxygen (O_2),

$$\begin{gathered} SF_6 + Si \rightarrow SiF_4 \\ O_2 + C \rightarrow CO_2 \\ O_2 + H \rightarrow H_2O. \end{gathered}$$

The most prevalent way is to force the Si to react with fluorine to a tetrafluorosilane (SiF_4) gas, the carbon with oxygen to carbon dioxide and the hydrogen with oxygen to water.

Typically BCB is structured either with hard [30, 31] or soft [32, 33] masks. It is also possible to use dielectrics such as Si_xO_y or Si_xN_y, as well as metal films are used as hard masks [34, 35, 29]. In the case of this dissertation, photo resists were used as etch masks.

In terms of etch selectivity [36] hard masks are superior compared to soft masks, but they need additional process steps to utilize them, e.g. the pattern has to be transferred from a lithography mask to the hard mask material. To remove a hard mask, e.g. SiO_2 and Si_3N_4 or metal hard masks, bases or dedicated mask strippers have to be used [28].
Dependencies of power, pressure and etch gas composition on the etch quality were evaluated in [34, 37, 38, 31, 36, 39, 32, 35, 30, 33, 40, 41, 29]. Typically used gases are SF_6 and CF_4 with O_2 at a $1:5$ and $1:4$ ratio, respectively. Etch rates of $0.25\,\mu m/min$ to $1.6\,\mu m/min$ can be achieved at pressures of 50 mTorr to 550 mTorr with a power of 50 W to 550 W. It was shown that with increased concentration of SF_6 to more than 10 % SF_6 in O_2, the etch rate decreases, since the most parts of the BCB consisting of carbon and hydrogen and needs oxygen to end up in a volatile composition [34, 30, 32]. The etch rate can be easily increased with higher pressure and higher power. In the case of the increased pressure, the anisotropy suffers and leads to an isotropic etch profile.
The standard process for the InP BCB dry etching was established on the RIE plasma etcher. With SF_6 (3 sccm) mixed with O_2 (7 sccm), with an RF power of 100 W and a pressure of 1 Pa. The negative bias is around 340 V and the etch rate is 60 nm/min. The standard etch process is shown in Fig.3.10. To compare new etch recipes with each other and with the standard process, all the samples were etched for $4.1\,\mu m$ (measured with the in-situ tool mentioned in chapter. 3.4.1). For the following description the plasma parameters will shown as:

$$SF_6 - O_2 - 3 - 7 - - 100 - 1$$

„1st gas “ - „2nd gas “- „1st gas flow [sccm] “- „2nd gas flow [sccm] “- „ICP power [W] “- „RF power [W] “- „pressure [Pa] “.

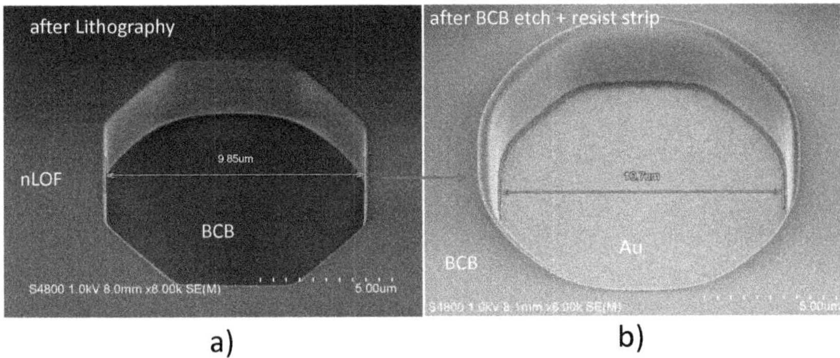

Figure 3.10 Pictures of the standard via process, a) scanning electron microscope (SEM) micrograph (30 ° tilted) of the resist after lithography, b) SEM micrograph (30 ° tilted) of the dry etched via with the standard RIE etcher, Parameters: SF_6 - O_2 - 3 - 7 - none - 100 - 1. at a bias of 340 V with an PA carrier.

The standard process has a widening of 0.7 μm compared to the designed layout (diameter of 10 μm), which is perfectly suitable for processing. In the following part of this chapter, the process development for a suitable and redundant etch process is described. The resist strip in the following development was performed on the ICP etcher in an O_2 (80 sccm) and He (10 sccm) ICP plasma with 400 W ICP and 0 W RIE power at 3 Pa process pressure [28]. In all processes the selectivity (between mask and BCB) was 1 : 1, since the BCB has a significant amount of organic material in the cross linked 3D network and therefore very similar to the resist mask.

3.4.3 BCB dry etching experiments before the equipment upgrade

At the beginning of the research the available etch recipes on the ICP etcher were evaluated. Those recipes were used for planarization etching and opening of big pad structures (> 100 μm) without a demand for high anisotropy but not used for small vias (< 10 μm). To keep the substrate and the carrier at the same temperature, a helium backside (He) carrier cooling was used at 7 °C (+/- 0.3 °C) [28].

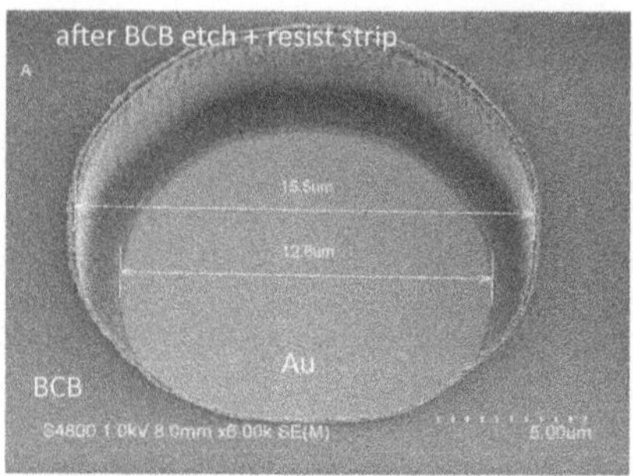

Figure 3.11 SEM micrograph (30 ° tilted) after dry etched via with the ICP etcher. Parameters: SF_6 - O_2 - 10 - 50 - 400 - 35 - 0.3 at a bias of 49 V with an aluminum (Al) carrier.

Fig.3.11 shows a post etch SEM image. It shows that a comparable isotropic etch profile results with high ICP power. Additionally, a higher gas flow is needed to maintain a consistent burning plasma, as the reactor is bigger than the RIE reactor. A total widening of 5.5 μm and 2.6 μm on top and at the bottom of the via, respectively, was observed. With such a widening it is not possible to process reliable interconnects in the InP process. Even with a precorrection in the lithography, the process cannot achieve reliable vias, not least because of the microloading effect. In this case, process development was inevitable for fast and reliable dry etching of BCB.

The first approach is to reduce the ICP power to achieve an anisotropic etch profile. Another approach is to reduce the high amount of oxygen, that supports the isotropic etch profile.

Figure 3.12 SEM micrograph (30 ° tilted) after dry etched via with the ICP etcher with an Al carrier. a) Parameters: SF_6 - O_2 - 10 - 50 - 200 - 35 - 0.3 at a bias of 85 V, b) Parameters: SF_6 - O_2 - 2 - 10 - 150 - 35 - 0.1 at a bias of 177 V.

As shown in Fig.3.12 a), reducing ICP power leads to a visible decrease of the isotropic profile. Nevertheless, the improvement was not sufficient, for present and future applications. In the subframe b) an approach towards the RIE process was tried. Here an additional improvement was observed, while decreasing the gas flows, the pressure and the ICP power. During the test for the new plasma parameters (without the test wafer), the border of the process windows was evaluated. With lower ICP power (below 150 W at the parameters of subframe b), the plasma was not stable. This was presumable due to the lower gas flows, pressure and the reduced ICP power, which is usually needed for an anisotropic process. With the results from the preceding experiments, another approach was tested. With the knowledge about the process boundaries for a consistent burning plasma, a higher gas flow and higher pressure are needed.

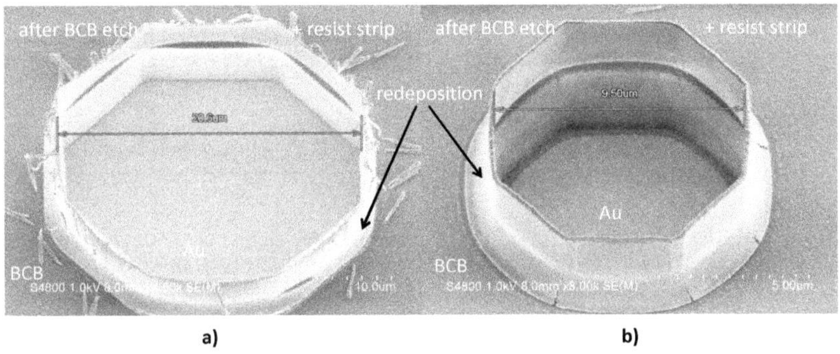

Figure 3.13 SEM micrograph (30 ° tilted) after dry etched via with the ICP etcher. Parameters: SF_6 - none - 40 - 0 - 0 - 100 - 1.2 at a bias of 250 V with an Al carrier, a) 20 μm via, b) 10 μm via.

As depicted in Fig.3.13, an anisotropic etch profile was successfully achieved with the ICP etcher. Unfortunately a significant redeposition after the etching was observed. In the subframe a) and b) are two pictures from one wafer (same experiment), which are showing a perfect reproduction of the resist sidewall shape. This redeposition was not clearly visible in the microscope, but in the SEM. Those structures were only observed where photo resist was opened and not on the surface of the plain BCB. Such redeposition is unwanted and due to the danger of collapsing into the hole and degrading or much more worse, destroying the interconnect. To understand the root cause, the wafer was examined by EDX equipment.

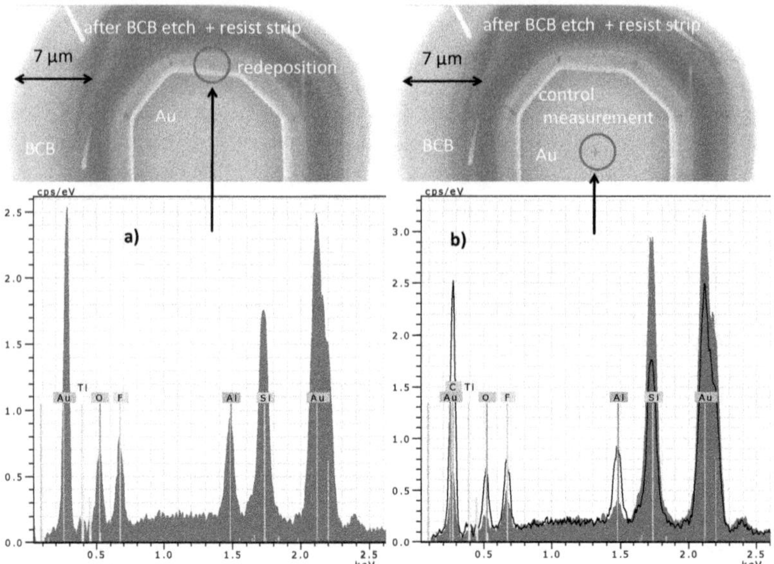

Figure 3.14 SEM micrographs and EDX spectra after dry etched via with the ICP etcher. Parameters: SF_6 - none - 40 - 0 - 0 - 100 - 1.2 at a bias of 250 V with an Al carrier, a) EDX measurement of the redeposition, b) control measurement outside from the redeposition.

As depicted in Fig.3.14, the redeposition consists presumably of a composition of fluorine, oxygen and aluminum. Possible solid compositions are AlF, Al_2O_3, AlO_xF_y. The origin of the Al is most likely from the wafer carrier, since it is completely made from Al or the reactor itself where all the walls are made of Al, which is unlikely due to the effect of the sheath region described in chap. 3.2.1. The fluorine origin is most likely from the gas mixture itself. The origin of the oxygen is not obvious, but the BCB itself has a oxygen binding between two Si atoms. Additionally the the oxygen and aluminum could be from the coupler plate beneath the ICP antenna or the carrier clamp, which are both made from aluminum oxide (Al_2O_3) [28].

Due to the results of the experiments and the outcome of the EDX spectra, the substrate carrier material was changed. Additionally, a hardware upgrade for the plasma matchbox needed to be evaluated, which enabled a reliable RIE process on the ICP reactor. The matchbox upgrade was a commercial product, supplied by the manufacturer (Sentech Instruments GmbH). As new carrier material, the polymer polyether ether ketone (PEEK) was chosen, due to its high melting point (> 340 °C) and its volatile by-products. The intention was to exclude the factor of the not volatile Al by-products, which up to this point were present with the Al carrier. As seen in Fig.3.15, a comparison on the RIE etcher with an polyamide (PA) carrier is shown. The process was set similar to the experiment shown in Fig.3.13 to compare the resulting redeposition. The specimen with the PA carrier showed no redeposition on the RIE reactor, therefore PEEK was chosen as carrier material for further investigations.

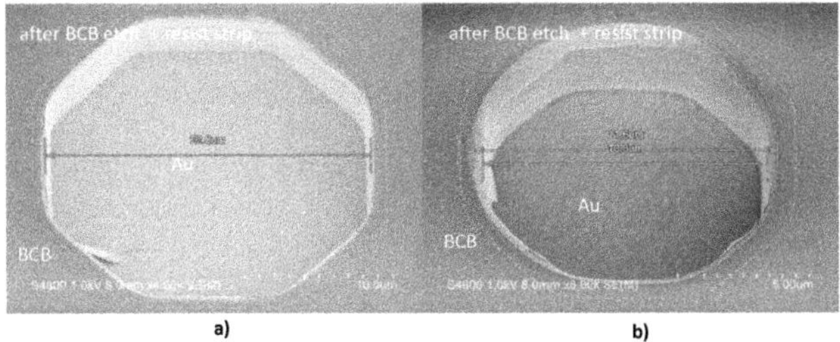

Figure 3.15 SEM micrographs of dry etched vias with the RIE etcher. Parameters: SF_6 - none - 40 - 0 - 0 - 150 - 1.0 at a bias of 370 V with a Pa carrier, a) 20 μm via, b) 10 μm via.

3.4.4 BCB dry etching experiments after the equipment upgrade

As the BCB manufacturer suggests [17] various gases and compositions, we used a gas ratio of O_2:SF_6 of 4:1, a pressure of around 1 Pa (see Table 3.1) and RIE power between 100 W and 200 W [28]. With increased RIE power, the etch rate increases, since more electrons facilitate more ions and radicals, higher bias and better anisotropy as seen in Tab. 3.1 and Fig.3.16. To avoid induced damage [27], the bias was monitored and kept at the level of the reference process (340 V). From this point also the new carrier (PEEK) was used to reduce the non-volatile Al by-products.

Table 3.1 Dry etch experiments after matchbox upgrade with the new carrier (PEEK)

SF_6 [sccm]	O_2 [sccm]	ICP [W]	RIE [W]	pressure [Pa]	etch rate [nm/min]	bias [V]
5	20	0	100	1	75	200
5	20	0	200	1	120	350
5	20	50	100	1	120	235
5	20	100	100	1	220	230
5	20	100	200	0.2	300	340
5	20	100	200	0.5	300	350
5	20	100	200	1	300	350

As shown in Tab. 3.1, the first recipe is on purpose very similar to the reference process. Due to the bigger reactor, the gas flows were increased. With a RIE power of 100 W, the etch rate of 75 nm/min was comparable to the reference process etch rate of 60 nm/min. To boost up the etch rate, further ICP power was added. Sufficient anisotropy was achieved by keeping the ICP:RIE power ratio below 1:2. The etch rate of 300 nm/min and comparable bias of 350 V were achieved with a gas flow ratio of 5 sccm (SF_6) : 20 sccm (O_2), 100 W ICP power and 200 W RIE power. Unexpectedly, the change of the substrate carrier did not result in a redeposition free etch picture, even with the PEEK carrier as seen in Fig.3.16.

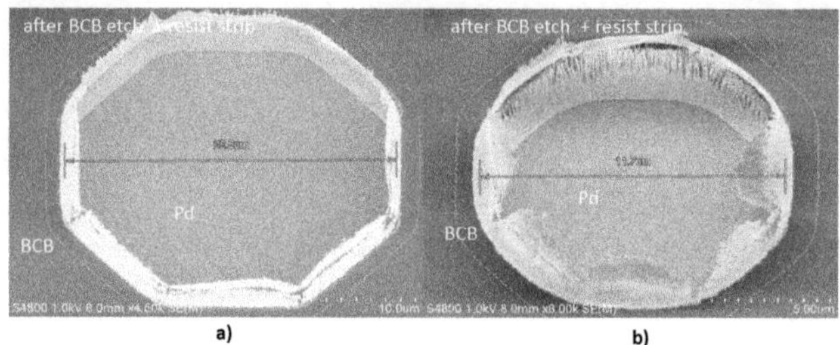

Figure 3.16 SEM micrographs of dry etched vias with the ICP etcher after matchbox upgrade. Parameters: SF_6 - O_2 - 5 - 20 - 100 - 200 - 1.0 at a bias of 340 V with a PEEK carrier, a) 20 μm via, b) 10 μm via.

An anisotropic etch recipe was chosen: SF_6 - O_2 - 5 - 20 - 100 - 200 - 1.0 at a bias of 340 V with a PEEK carrier and an etch rate of 300 nm/min. The remaining problem with the redeposition was in the main focus, after a suitable etch recipe was found. Another experiment, to see a direct dependency between different carriers, was performed.

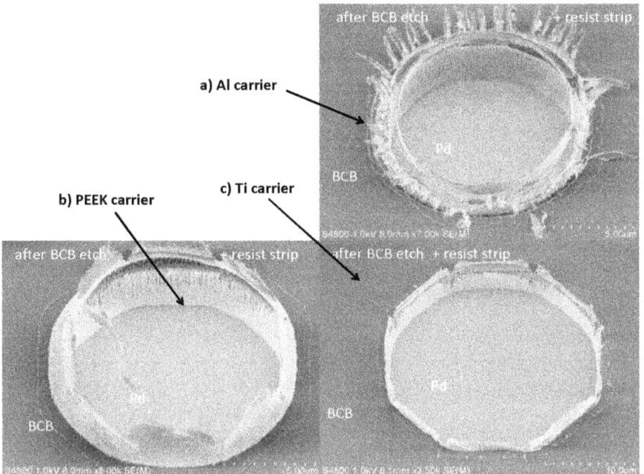

Figure 3.17 SEM micrographs of dry etched vias with the ICP etcher. Parameters: SF_6 - O_2 - 5 - 20 - 100 - 200 - 1.0 at a bias of 340 V with different substrate carrier, a) Al carrier, b) PEEK carrier, c) Ti carrier. Based on [28]

As depicted in Fig.3.17, the redeposition is visible with all used carrier. The shape and amount of the by-product is different, but present. The etch rate and bias were also affected by utilizing different carriers, as seen in Tab: 3.2.

Table 3.2 Etch rate and bias as a function of different substrate carrier

carrier	etch rate [nm/min]	bias [V]
Al carrier	440	430
PEEK carrier	300	350
Ti carrier	384	430

The lower etch rate with the PEEK carrier can be explained, due to its polymer components, which therefore also consumes the reactive species. The total area which can be etched is increased and depletes the ions and radicals. Both metallic carriers have a higher etch rate and higher bias than the PEEK carrier. This can be explained with stronger fields at the electrode, which leads in the sheath region at the electrode to higher bias and simultaneously stronger acceleration of the ions. The lower etch rate of the Ti carrier compared with the Al carrier is the same mechanism as with the PEEK carrier. The Ti is etched in SF_6 [42] and leads to a depletion of the reactive species. The Al carrier remains mostly unetched. Apparently, Ti has a less profound redeposition compared to the other carrier but is still present. The formation of the redeposition with the Al carrier was unexplainable different

due to the spikes seen in Fig.3.17 subframe a). Presumably the different shapes of the formed residues depend on the electromagnetic fields which are unique for the given materials.
To understand the origin of the still present redeposition, another EDX analysis was done.

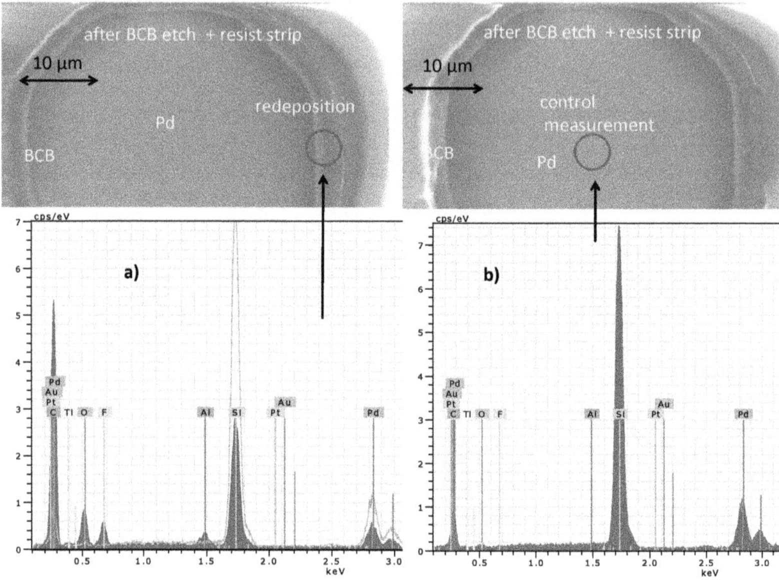

Figure 3.18 SEM micrographs after dry etching, resist stripping and cleaning with different chemicals, and EDX spectra after dry etched via with the ICP etcher. Parameters: SF_6 - O_2 - 5 - 20 - 100 - 200 - 1.0 at a bias of 340 V with PEEK substrate carrier, a) EDX measurement of the redeposition, b) control measurement outside from the redeposition.

As depicted in Fig.3.18, the EDX spectra showed the same elements as in Fig.3.14 with Al as carrier material with massive redeposition. It was unexpected, but the change of the carrier material did not result in a aluminum, oxygen and fluorine free redeposition. For further experiments the PEEK carrier was used, since its composition is etchable in the chosen process parameters. To find some evidence of the origin, the reactor was opened. The opened reactor is visible in the Fig.3.19.

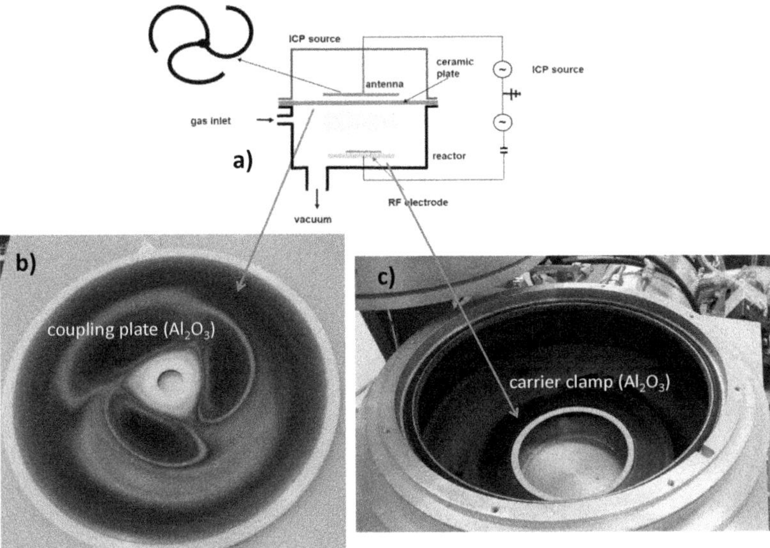

Figure 3.19 a) Principal ICP schematic construction. Based on Fig.3.7, b) image of the coupling plate beneath the ICP antenna, c) image of the carrier clap and the bottom part of the ICP etcher

As seen in subframe b) of Fig.3.19, the coupling plate (Al_2O_3) is covered with non-volatile by-products as well es the reactor itself, seen in subframe c). The middle shape of the coupling plate is bright and presumably sputtered with some reactive species. Is is also visible that the shape of the ICP antenna is patterned into the coupling plate. The antenna has a massive ring of copper in the middle of the coupling plate, which creates a negative electric field like at the bottom electrode, where positively charged ions can be accelerated to. However, this is most likely the origin of the aluminum and presumably of the oxygen in the redeposition. The fluorine is most likely from the reactive plasma itself. With this information further process development to circumvent the redeposition in the plasma process was stopped. Ways to suppress such redeposition within the construction of the plasma equipment are explained in chap. 3.5. The formation of the redeposition is most likely formed by sputtered Al_2O_3 from the coupling plate underneath the antenna. The AlF_xO_y based residues coating the whole wafer conformally (typical for sputter process). The horizontal area is continuously cleaned by the incoming ions, but the vertical surfaces are not cleaned, due to the anisotropic etch regime. In the middle the ICP configuration, where the ICP power is fed into the antenna, a

discontinuity occurs. This creates a field that is perpendicular to the coupling plate, which leads to the same self bias effect, which in was mentioned chap. 3.2.1.
Since the avoidance of the redeposition seemed to be unlikely, due to the construction of the etcher, it was decided to seek for a clean and selective way to remove the redeposition after the plasma process.

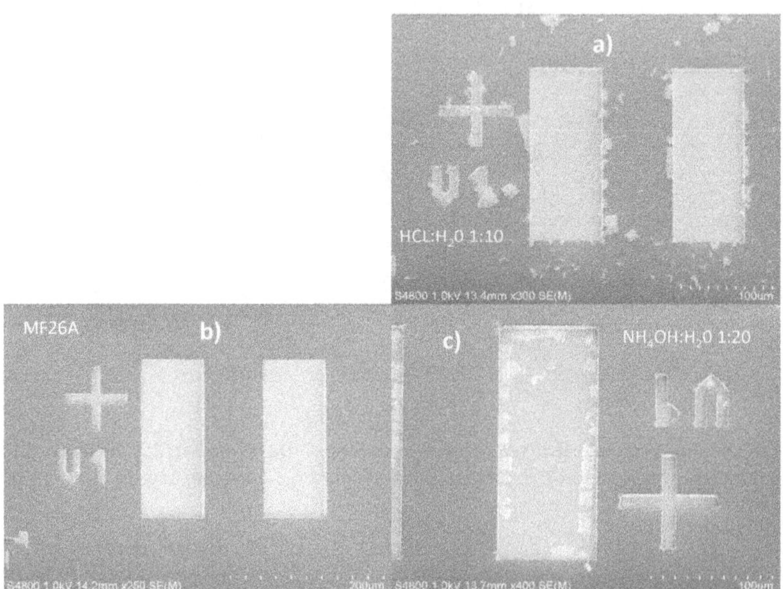

Figure 3.20 SEM micrographs after dry etching, resist stripping and cleaning with different chemicals, Parameters: SF_6 - O_2 - 5 - 20 - 100 - 200 - 1.0 at a bias of 340 V with PEEK substrate carrier, a) HCl (37.5 %) in deionized water (DI) water with a 1:10 ratio for 30 sec, b) tetramethylammonium hydroxide (TMAH) developer for 60 sec, c) NH_3 (25 %) in DI water with a 1:20 ratio for 30 sec.

Published research reports the removal of redeposition at the via bottom after BCB interconnect etch [37] with certain chemicals. Their removal was performed with commercially available stripping agent. However, due to the amphoteric property of aluminum, the removal the residues with acids and bases was tried. The chemical agent must be compatible with the InP process, therefore aggressive chemical agents such as potassium hydroxide, hydrofluoric acid, or piranha solution were not an option. Those acids would damage the underlying layer with the active devices. Chemicals which are part of the process and are less harmful were used: ammonia (NH_3) (25 %) in DI water with a 1:20 ratio, hydrochloric acid (HCl) (37.5 %)

in DI water with a 1:10 ratio and TMAH developer with 0.26 normality (MF-26A, Dow Chemical, USA) [28].
In Fig.3.20, the results after the usage of different chemicals to remove the residues are shown. $HCl:H_2O$ and $NH_3:H_2O$ were not able to remove the redeposition in the given dilution. The best result was achieved with MF-26A. It is known, that with etch rates of 50 nm/min to 100 nm/min, TMAH-based developers can etch aluminum [43]. With the MF-26A a suitable process to remove the redeposition after plasma etching was found. The MF-26A is often used in the process of resist development, which leads to the conclusion, that it is not harming the MMICs or BCB stack.
With the MF-26A, the substrates of Fig.3.17 were cleaned, to see if the assumption concerning the origin of the residues is true (coupling plate and not carrier depended). The results are shown in Fig.3.21.

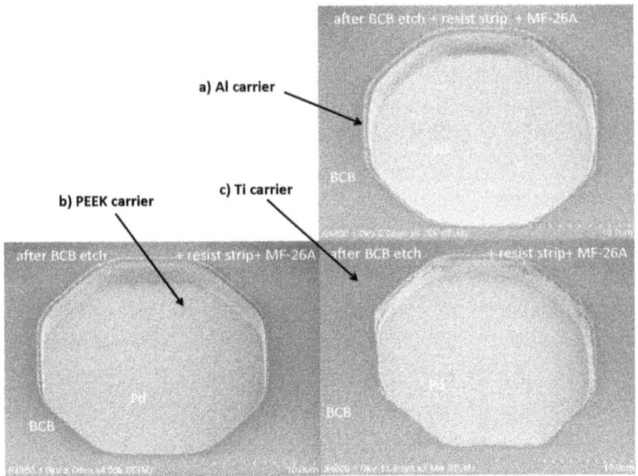

Figure 3.21 SEM micrographs of dry etched vias with the ICP etcher. Parameters: SF_6 - O_2 - 5 - 20 - 100 - 200 - 1.0 at a bias of 340 V with different substrate carrier and cleaned with MF-26A, a) Al carrier, b) PEEK carrier, c) Ti carrier. Based on [28]

With a suitable etch process and a method to remove the residues, a further experiment was done to see a dependency of pressure on the anisotropy. While a definite dependency of pressure and anisotropy was reported [30], in this work such trend in the RIE-like process in the ICP etcher was not osereved. To guarantee a stable plasma, the process window concerning the pressure was limited between 0.1 Pa and 1 Pa. The results are shown in Fig.3.21. The minimal variation of the pressure (mean free path), resulted in no difference of anisotropy.

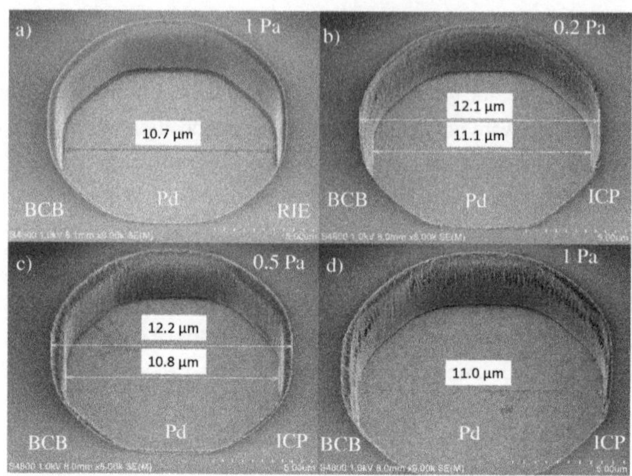

Figure 3.22 SEM micrographs after dry etching, resist stripping and cleaning with MF-26A, a) after RIE process as in Fig.3.10, For b), c) and d) processed with SF_6 - O_2 - 5 - 20 - 100 - 200 - X at a bias of 340 V with PEEK substrate carrier and 0.2, 0.5 and 1 Pa, respectively. Based on [28].

With an etch recipe and a method to clean the wafer after plasma processing, the uniformity of the plasma process was investigated. Two substrates were coated with 2.6 µm BCB without an etch mask. The thickness of both substrates was measured with an ellipsometer (80 points over the 3 inch wafer area), see Table 3.3. On both wafers, the BCB was etched circa 1.25 µm. The results showing less than a threefold standard deviation (SD) increase after etching, over the whole wafer of less than 20 nm for wafer 1 and less than 10 nm for wafer 2, respectively. Such deviation corresponds to 1 % of the total etched depth [28]. With such an uniform etch process, the new recipe is suitable for the InP HBT process.

Table 3.3 BCB thickness before and after etching with the newly developed ICP recipe, 80 points ellipsometer measurement over full 3" wafer

	pre etch wafer1	post etch wafer1	pre etch wafer2	post etch wafer2
mean [nm]	2729	1479	2585	1345
SD [nm]	6	12	7	10

3.5 Methods to suppress redeposition

As already mentioned in chap. 3.4.4, there are methods to suppress the formation of the redeposition. Both solutions were considered and one of it evaluated. Due to the FBHs clean room is shared with other departments, many processes have to run reliable on this ICP etcher, to not compromise parallel projects. Therefore, none of both methods were implemented to avoid the compromising the other experiments and departments at the FBH.

Figure 3.23 Image of the suspended quartz coupling plate beneath the ICP antenna. Based on [44]

The first method would require an additional suspended quartz coupling plate. This would replace the Al_2O_3 coupling plate. In such a configuration the material would be etchable in the SF_6 plasma. Consequently, the elements of the quartz plate would form volatile by-products and not coating the substrate [42]. With such an upgrade, all existing etch recipes would need an adjustment, which is a disadvantage for the massive amount of parallel projects on this equipment.

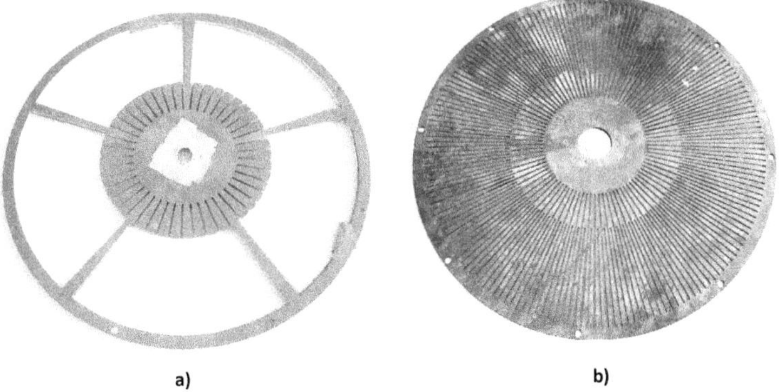

Figure 3.24 Image of the Faraday shielding for mounting beneath the coupling plate, a) partially closed Faraday shielding, b) fully closed Faraday shielding. Based on [44].

A different approach to suppress the redeposition is an implemented Faraday shielding [45]. This would be also a suspended upgrade underneath the coupling plate. In such an upgrade, the metal needs to be grounded, to repel the incoming electrons from the sheath region. This would reduce the self-bias effect and create a barrier such as with the grounded reactor wall (described in chap. 3.2.1). This kind of experiment was performed at the FBH by a colleague (Dr. R.S. Unger) after the experiments of the author of this dissertation.

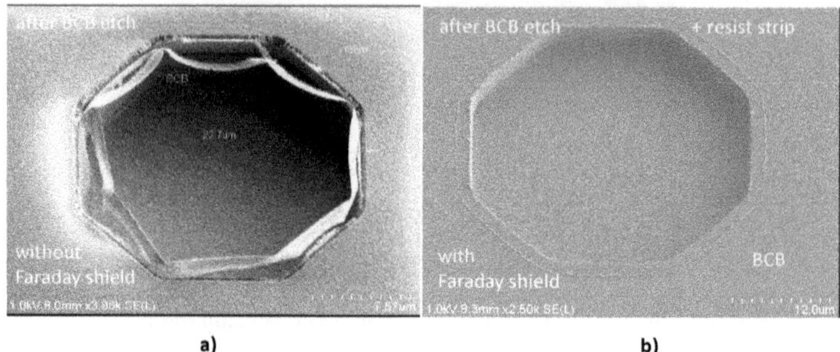

a) b)

Figure 3.25 SEM micrographs after dry etching, resist stripping with a PEEK substrate carrier, a) BCB etching without Faraday shield SF_6 - O_2 - 10- 50 - 100 - 200 - 1 , b) BCB etching with a partially filled Faraday shield SF_6 - O_2 - 10- 50 - 300 - 200 - 1, Based on [44].

The results of the Faraday shield experiment are shown in Fig.3.25. As seen, the utilization of the Faraday shield resulted in a residues free etch process. The ICP power needed to be increased, since the shielding absorbs a huge amount of the ICP power. Consequently, this is the reason for not using the shielding, due to the fact, that all recipes would need an adjustment and would compromise the parallel projects [44].

3.6 Implementation into the InP HBT process

The newly developed process was implemented in the InP process. To verify the new process with electrical data, process control monitoring (PCM) was analyzed. Therefore, two wafers each, old process (RIE) and new process (ICP), were compared. Interconnects from G2 to G1 (V2) and G1 to GD (V1G) were evaluated, as seen in Fig.3.26 in subframe a). In subframe b) in Fig.3.26 an example of a 32 point PCM measurement is shown.

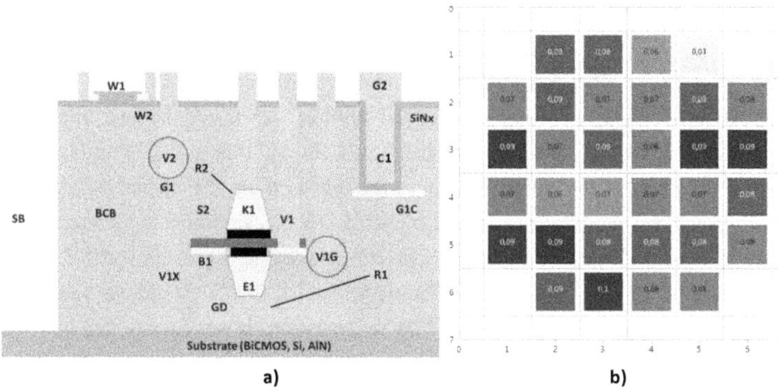

Figure 3.26 a) illustration of the location of the two etch steps for PCM measurement, b) example of a 32 point PCM measurement with the old (RIE) process.

The results of the PCM evaluation is shown in Tab. 3.4. DC measurements were performed on a Karl Süss PA200 DC prober with 32 point measurement for each data point and wafer. Two conclusions can be drawn from the PCM measurements. First, the data points are not far apart and the difference between each wafer can be seen as process deviation depended. Second, the data points of the newly developed process are slightly better than with the old process. The evaluated vias are octagonal interconnects with a diameter of 8 μm. Consequently, the new developed process can be used in the InP DHBT process and shows better electrical performance at maintained anisotropy with a fivefold etch rate, compared with the old process.

Table 3.4 PCM measurement with the old and new plasma recipe on fully processed InP DHBT wafers

	via	min [Ω]	max [Ω]	mean [Ω]	S.D. [Ω]
RIE Wafer 1	V2	0,01	0,05	0,03	0,009
RIE Wafer 2	V2	0,007	0,04	0,03	0,007
ICP Wafer 1	V2	0,006	0,02	0,02	0,004
ICP Wafer 2	V2	0,007	0,02	0,02	0,005
RIE Wafer 1	V1G	0,03	0,1	0,08	0,01
RIE Wafer 2	V1G	0,02	0,09	0,07	0,02
ICP Wafer 1	V1G	0,02	0,02	0,02	0,002
ICP Wafer 2	V1G	0,01	0,04	0,02	0,009

3.7 Chapter 3 summary

This chapter presents the development of a novel etch process for BCB dry etching. The necessity for a redundant and faster process was fulfilled. A fivefold increase in etch rate was achieved at maintained bias and anisotropy. It was shown, that no RIE dominated process with a standard configuration (matchbox) was feasible with the ICP etcher. A major problem with residues was found during the experiments. EDX measurements indicated an AlF_xO_y composition of the residues. After the matchbox upgrade a new process was developed with a suitable bias, anisotropy and etchrate. The main focus of the investigations was transferred after the matchbox upgrade to the origin of the redeposition. It was shown, that with different substrate carriers the residues are still present, but in different shapes and quantity. The origin of the redeposition was traced back to the coupling plate made of Al_2O_3. As a workaround, a method was shown to clean the substrates after the plasma processing. The TMAH-based MF-26A developer was the best solution, leaving no residues behind without harming the MMIC stack. Within a small variation of the reactor pressure, no impact on the anisotropy was found. The new developed process indicated a suitable etch uniformity across the whole 3 inch wafer. Also, a method to suppress the redeposition formation was shown. At the end, the implementation of the ICP process was compared with the standard RIE process. The new process has slightly better electrical performance than the RIE process. In sum, the shown plasma etch process (ICP) was successfully implemented into the InP processing. The experience in suppressing the redeposition was considered in the purchase of new plasma etch equipment.

Chapter 4

Nickel chrome thin film resistors

NiCr resistors were introduced in chapter 1 as the second part of this dissertation. A resistor has numerous of applications in the electronic world, which will be described in section 4.1 of this chapter. Until the NiCr resistors have been developed and implemented in this work into the design kit for the InP DHBTs, only low resistivity resistors with significant parasitic elements could be used in the InP process at the FBH.

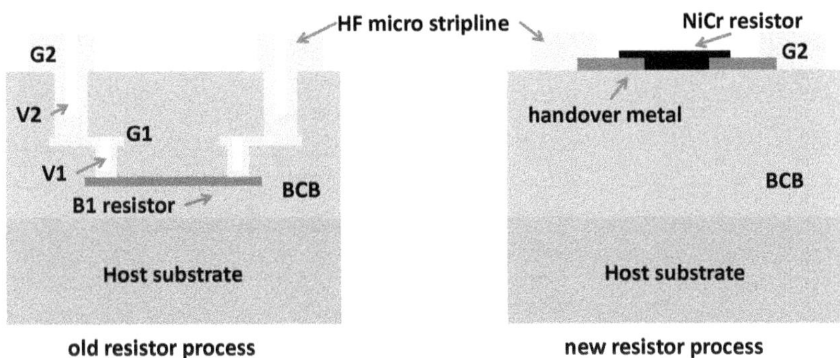

Figure 4.1 Comparison between the predecessor (left) and the new (right) process scheme for the resistor.

As shown in Fig, 4.1, the predecessor resistor process utilizes the base metal as resistive area. This process scheme has the benefit of utilization of an existing process metal where no additional dedicated resistor lithography layer is needed. The drawbacks of the previous process are low sheet resistance (0.5 $\Omega/\square$) which leads to huge area demands and the parasitic behavior. Those resistors are useful only for special purposes. The new resistor process scheme has a high sheet resistance (25 $\Omega/\square$) and neglectable parasitic elements, which promises to be a suitable broadband resistor for most applications in a circuit. In this chapter, the fundamentals, thermal and electrical simulations, process development, DC and HF measurements of the NiCr resistors will be shown.

4.1 Fundamentals of thin film resistors in electronics

Thin film resistors (TFR) have numerous applications in integrated circuits. Resistors are necessary to adjust biasing of transistors e.g. at the base of an bipolar transistor to enable a high current flow through the collector. Implemented resistors in series or in parallel, can reduce the voltage and simultaneously adjust the current. Furthermore, a resistor can be used together with a capacitor as a timing element in DC applications.
Additionally, for high frequency applications, the resistors can be utilized to serve as terminations, attenuators, Wilkinson power dividers, filters and stability networks [46].

$$\mathrm{R} = \frac{\mathrm{U}}{\mathrm{I}}; \quad \mathrm{I} = \frac{\mathrm{U}}{\mathrm{R}}; \quad \mathrm{U} = \mathrm{RI}. \tag{4.1}$$

For a homogenous resistive element (e.g. TFR), R can be calculated from the quadratic sum of blocks (squares), if the resistance of a square block is known as shown in Fig.4.2. The resistance of one block (square) is described as sheet resistance $\mathrm{R_S}$.

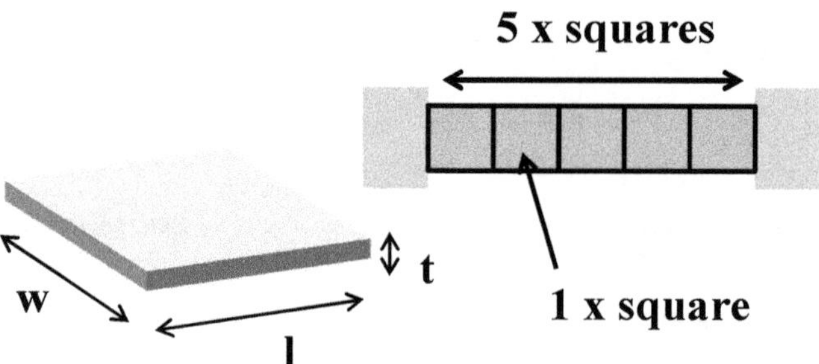

Figure 4.2 Simple thin film resistor geometry. Based on [16].

To calculate $\mathrm{R_S}$, the $\ell = \mathrm{w}$ needs to be equal [47]. By using a quadratic geometry ($\ell = \mathrm{w}$), $\mathrm{R_s}$ is inversely proportional to the thickness t in the resistive material with a specific resistivity ρ.

$$\mathrm{R_S} = \frac{\rho}{t}. \tag{4.2}$$

For high frequency operations, the skin effect needs to be considered. The skin effect describes a dependency between the effective area (e.g. cross section of a wire) in which the current flows and the frequency of the applied alternating current. With increasing frequency, the effective volume that serves as conducting material shrinks. Subsequently, the resistance increases with rising frequency, since the resistive area is reciprocal connected to the resistance [48].

$$\mathrm{d} = \frac{1}{\sqrt{\pi f \mu \sigma}}. \tag{4.3}$$

As seen in Eq. 4.3, the skin depth d depends reciprocal on square root of π, frequency f, permeability μ and the conductivity σ. Subsequently, the effective area that contributes to flowing current, decreases with increasing frequency [48].

With typical material parameters, the skin depth in NiCr results in several micrometer. Due to the shallow NiCr TFR thickness of maximum hundred nanometer, the skin effect is not affecting the device performance. At higher frequency, the parasitic inductance and capacitance (e.g. TFRs area to ground) have the most impact on the HF parameter.

An important method to calculate and track the contact resistance of metal to metal and also metal to semiconductor is the transmission line method (TLM). Within this method, the contact resistance R_c [Ω], the sheet resistance R_s [$\Omega/\square$] the transfer legth L_T [μm] and the specific contact resistance ρ_c [Ωcm] can be calculated. The transfer length is the average distance that an electron travels beneath the contact until it flows up in the contact. The specific contact resistance is material depended and describes the characteristic contact resistance between two given materials. The contact resistance is the absolute resistance that adds to a resistive line between to layers [49].

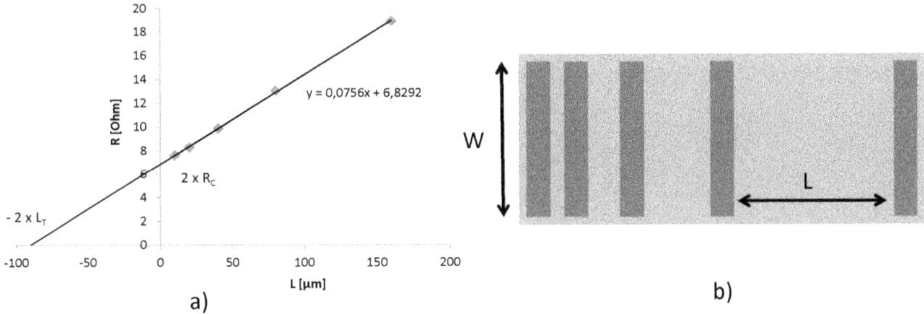

Figure 4.3 Example an TLM measurement and layout. a) example TLM measurement data, b) TLM layout. Based on [49].

In Fig.4.3 an example of a TLM measurement is shown together with a TLM layout. In this contact resistance extraction, a few similar shaped contacts are applied on a material. The distance needs to be variegate to extract a slope of such linear line. In this case (Fig.4.3 subframe a) five distances L (μm) are shown with the same width W (100 μ). By measuring the resistance between each line, a slope can be calculated. With the slope the actual and absolute contact resistance can be calculated.

$$\mathrm{R_c} = \frac{\text{trend line intersection with y}}{2}. \tag{4.4}$$

The intersection of the trend line with the y-axis is the twofold of the contact resistance $\mathrm{R_c}$. The sheet resistance can also be calculated by this method.

$$\mathrm{R_s} = slope\ x\ W. \tag{4.5}$$

By multiplying the slope with the width of the lines, the sheet resistance $\mathrm{R_s}$ can be calculated. The transfer length can now be calculated by using $\mathrm{R_s}$, $\mathrm{R_c}$ and W.

$$\mathrm{L_T} = \frac{\mathrm{R_c} W}{\mathrm{R_s}}. \tag{4.6}$$

The specific contact resistance ρ_c can now be calculated.

$$\rho_c = \mathrm{R_c L_T} W. \tag{4.7}$$

Two further important extraction methods for the semiconductor technology are the Kelvin sensing and the van der Pauw method. The Kelvin sensing is used to measure the absolute contact resistance between two layers. The van der Pauw method is utilized to measure the sheet resistance of one sheet. Both values, $\mathrm{R_c}$ and $\mathrm{R_s}$, can be extracted in the TLM calculation but usually the Kelvin and the van der Pauw methods are implemented for redundancy and due to the low area requirement on a chip.

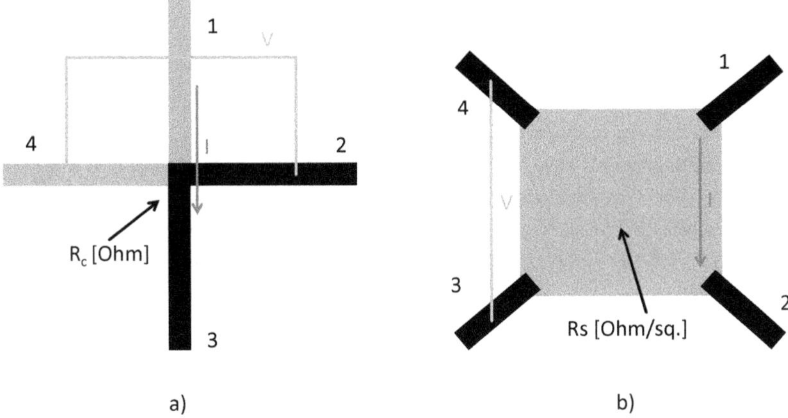

Figure 4.4 Kelvin sensing and van der Pauw method layout. a) Kelvin sensing, b) van der Pauw method (VDP). NiCr shown as grey and contact metal as black. Based on [50].

In Fig.4.4, a layout example of a Kelvin sensing and of the van der Pauw method is shown. The Kelvin sensing is a simple structure to calculate the contact resistance R_c. In the Kelvin sensing method a current is forced between port 1 and 3 (Fig.4.4 subframe a)) and the voltage is sensed between ports 2 and 4. Here, only the voltage drop in the middle of the cross is significant for the resistance between both layers. In the van der Pauw method, a current between the adjacent ports (e.g. 1 and 2 in subframe b)) is forced and the voltage is sensed at the opposite side (in this case 3 and 4).

$$R_{vdp} = \frac{V_{sense}}{I_{force}}. \tag{4.8}$$

The measurement is usually repeated four times to calculate a reliable average value. The R_{vdp} is the quotient of the voltage between (in this case) port 3 and 4 and the current between port 1 and 2. With the R_{vdp} the R_s can be calculated.

$$R_s = \frac{\pi R_{vdp}}{\ln 2}. \tag{4.9}$$

A simplified formula (Eq. 4.9) is commonly used to calculated the R_s with the van der Pauw method [50].

4.2 Process development

Apart from implanted or polysilicon resistors, commonly used materials for thin film resistors are tantalum nitride (TaN), titanium nitride, silicon cromium, tantalum, nickel, chrome, copper nickel and nickel chrome [47]. The most utilized materials for MMICs are TaN and NiCr [51]. Both materials have outstanding properties. However, Ni(80 %)Cr(20 %) related to weight percent, in comparison to the above mentioned materials, shows the best thermal long-term stability of $< 0.02\,\%$ after 1000 h at 150 °C and the lowest temperature coefficient (TCR) of $\alpha = 120$ [ppm/K] [16]. Therefore, NiCr was chosen as the resistive material for the experiments.

4.2.1 Methods and Tools

NiCr and SiNx deposition was performed by using a sputtering tool. The sputtering process relies on the bombardment of a surface (sputter target) with accelerated particles. By hitting the sputter target, atoms are removed from the target and are deposited on the desired area (substrate) beneath. Such an accelerated particle can be an ion or an electron. Due to the comparably high energy, ions are used as the incident particle, due to the higher mass of ions compared to electrons. Argon (Ar) is the most common element, due to its inert and ionization properties [24].
Depending on which element is sputtered, the sputter yields differ. The sputter yield describes the measurable amount of ejected atoms by using the same ion energy and ion density. This phenomenon leads to an effect, especially within the NiCr sputtering, of a shifting composition ratio between the sputter target and the deposited layer. Nickel has a sputter yield of 1.9 compared to 1.7 for chrome at an ion energy of 1000 eV. This effect leads to a slight shift to chrome, which is not exactly Ni(80 %)Cr(20 %) after the deposition [24].

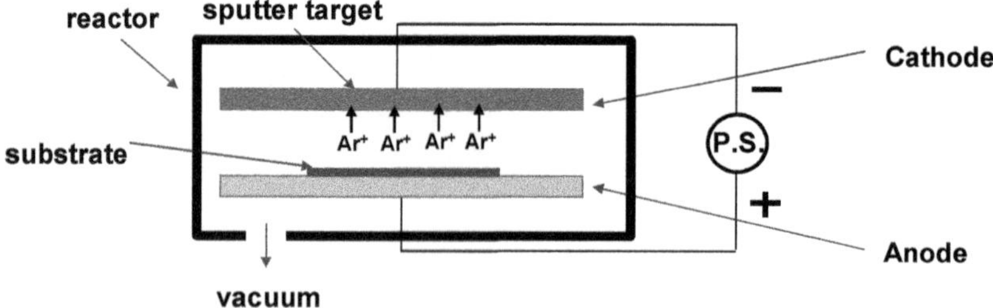

Figure 4.5 Illustration of the sputter process in a DC sputter reactor. Based on [24].

As seen in Fig.4.5, the DC sputter reactor is similarly constructed as the plasma reactor described in Chap. 3.3. The sputter target with the desired material is connected to the negative pole of the power source to attract the positive charged Ar ions. The ejected atoms of the sputter target are deposited on the substrate. Such a DC sputter reactor is suitable and used for metals as target material. For dielectrics like SiNx, the magnetron sputtering technique (coil placed above the sputter target) is used. With a magnetron, an magnetic field is induced to lead the electrons (which are necessary to ionize the Ar atoms) on longer paths, which increase the probability of hitting an Ar atom. With more ionized Ar ions in such a magnetron reactor, the damped target (e.g. SiNx) is now capable to eject more atoms as compared with a DC setup with a dielectric target material [24].
It is well known, that by changing the deposition parameters such as argon flow rate, power, chamber pressure and distance between target and sample, the stress, grain size, resistivity and TCR of NiCr can be adjusted [52].
As described in Chap.4.2, the NiCr resistive layers are deposited by sputtering. NiCr was sputtered using a VON ARDENNE sputter system CS 730. All experiments were processed with a target-to-substrate distance of 65 mm. The process pressure was fixed at 8 x 10^{-3} hPa with an Argon flow of 50 sccm. The NiCr layers were sputtered at an RF power of 400 W power, which resulted in a 376 V bias voltage. On top of the NiCr, a 50 nm thick layer of SiN_x was sputtered to prevent the NiCr from oxidization at atmosphere [53]. This SiN_x was sputtered with the same system without exposing the NiCr to atmosphere. The SiN_x sputtering was deposited with a process gas of 40 sccm argon and 4 sccm nitrogen and a power of 100 W in RF mode at a pressure of 5 x 10^{-3} hPa [16].
Further used equipment:

- scanning electron microscopy (Hitachi S4800)
- plasma etcher SI 500 from Sentech Instruments GmbH
- i-line step-and-repeat lithography tool (Nikon i12)
- dedicated BCB curing oven (YES-PB6-2PCP)
- electroplating with a commercial system from Mikro- und Oberflächen Technik GmbH
- mechanical profiler P10 and P16 from KLA
- metal evaporator for connections LH560UHV from Leybold
- DC measurement tool PA200 for (dedicated for process control measurements) from Karl Süss PA200

- four point DC measurement tool Veeco AP-150 probe

4.2.2 Nickel chrome technology development

To verify the NiCr sputter target, a NiCr layer with a thickness of 44 nm was deposited on double side polished square glass samples (1600 mm^2). The resulting specific resistance of the NiCr sheet was 110 $\mu\Omega$cm [16]. Such specific resistivity is a typical value for Ni(80 %)Cr(20 %) [54, 52, 47].

Since the NiCr resistors are intended to be implemented in the InP MMIC process, the particularities of this integration have to be considered. At the beginning of Chap. 4, the applications of resistors for DC and HF are described. Additionally, the necessity for a TFR integrated on the top surface (BCB) in the InP MMIC process is shown. For a good thermal connection, the NiCr resistors should be placed as closely as possible or directly on the host substrate for DC applications. To meet the demands for HF applications, the NiCr TRFs should be placed on the same layer as the signal leading HF microstrip line (G2 interconnection metal in the InP process). Here, a tradeoff in the middle of the stack would not help either application, since it would have parasitic connections for the HF and still a high thermal resistance due to the BCB between the TFR and the host substrate. Therefore, it was decided to place the NiCr resistor on the top surface and limit the allowed current sustainability. This has the drawback of a higher thermal resistance due to the low thermal conductivity of BCB but allows good broadband characteristics due to the avoidance of parasitic HF elements.

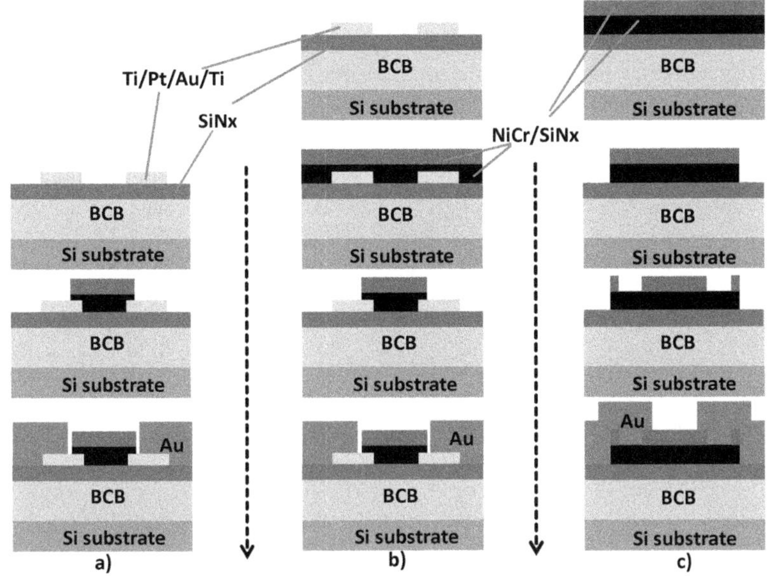

Figure 4.6 Experiment schemes for the TS process implementation, a) bottom connection with lift-off (BCL), b) bottom connection with wet etch (BCW), c) top connection with wet etch (TCW). Based on [16].

In Fig.4.6, the experimental setup for the NiCr integration is shown. Subframes a), b) and c) outline three different integration approaches. Approach a), is integrated with NiCr connected from the bottom and structured by lift-off. Approach b) is mostly similar to a), except for the NiCr structuring, which is done with SiNx as an etch mask in a wet etch process. In approach c) the NiCr is structured by wet etching as in b), but the connection is established from the top to the electroplated interconnection metal.

All integration experiments were performed on full 3 inch silicon wafers $< 100 >$, which were undoped and single side polished. To ensure the full comparability to the actual process, 9 µm of BCB (bottom) with 200 nm SiN_x on top was applied on the silicon wafers. 4.5 µm thick electroplated interconnection gold (Au) metal was used as in the actual process. As plating base, 30 nm titanium tungsten (TiW) at the bottom and 100 nm (Au) on the top was used. TiW is needed for adhesion to the SiNx, since the precious Au metal shows poor adhesion. Additionally, the TiW has the benefit of etchability by hydrogen peroxide (H_2O_2) which is not etching SiNx. In the case of approach a) and b) a handover metal is needed to connect the electroplated gold with the NiCr. The common way is to deposit the NiCr on top of the interconnection metal. In this case, this would lead to a reliability risk, due to the significantly thicker gold layer (4.5 µm). To avoid cracks at the gold edge transition, the handover metal was implemented. The handover metal (from bottom to top) contains 30 nm titanium, 20 nm platinum, 100 nm gold and 30 nm titanium [16].

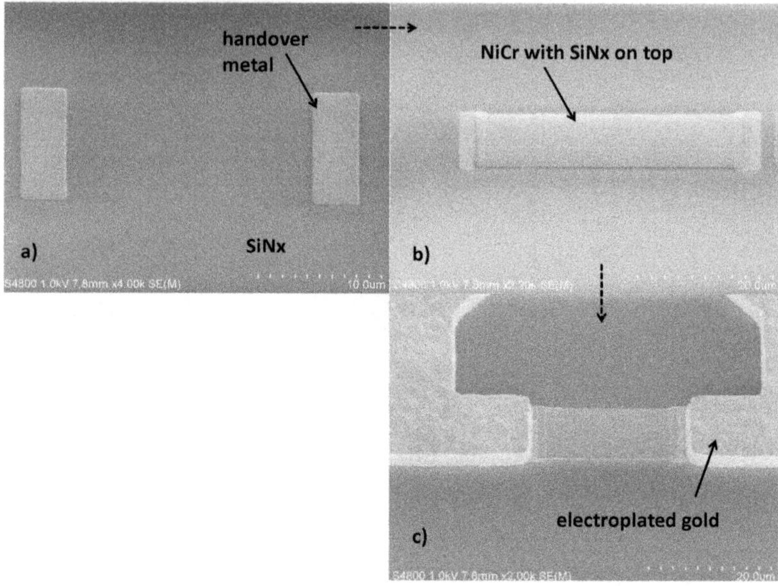

Figure 4.7 SEM micrographs of the bottom connection with lift-off (BCL) TFRs fabrication, a) handover metal after lift-off structuring, b) NiCr applied on the handover metal by lift-off, c) electroplated gold connected to handover metal.

In Fig.4.7, the processing of the BCL method is shown. Here, the handover metal is deposited by the lift-off technique. Therefore, preferably a negative photo resist is applied on the wafer and structured such that an undercut is yielded in the resist opening. After the photo resist, the metal is applied by sputtering or electron beam evaporation. With the metal on top of the resist, the area apart from the desired resistor or handover structure is lifted with the solvent N-Methyl-2-pyrrolidone (NMP). For the BCL experiments an AZ nLOF 2035 (MicroChemicals GmbH) for handover metal and an AZ 5214E for the resistor with a thickness of 1.7 µm and 2.2 µm, respectively was used. For electroplating (4.5 µm gold), an AZ nLOF 2035 with a thickness of 5.5 µm were used. Besides from adjusting the lithography parameter, the processing with the BCL method was fabricated without complications.

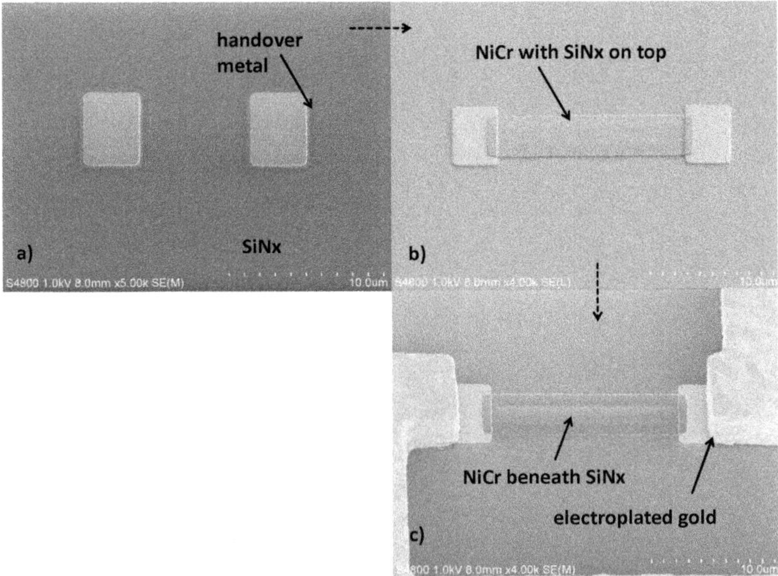

Figure 4.8 SEM micrographs of the bottom connection with wet etch (BCW) TFRs fabrication, a) handover metal after lift-off structuring, b) NiCr structured by wet etching, c) electroplated gold connected to handover metal.

In Fig.4.8, the processing of the BCW method is shown. The handover metal was applied similar to the BCL method, as described above. For NiCr structuring, the whole wafer was deposited with NiCr and SiNx as seen in Fig.4.6 in the same sputter tool. For structuring a positive photo resist (AZ 1518) was applied on the desired areas for the TFRs. With the photo resist, the SiNx underneath (SiNx above the NiCr) was structured by plasma etching. SiNx plasma etching was performed with a mixture SF6 (20 sccm) and He (10 sccm) with an ICP power of 300 W and RIE power of 50 W at a pressure of 0.3 Pa with a resulting bias of 100 V. With the structured SiNx etch mask, the photo resist was removed for NiCr wet etching. The NiCr was structured by Chrome Etch 18 (distributed by micro resist technology GmbH in Berlin, Germany) with an etching time of 90 sec at 20 °C. The interconnecting metal was similar performed as with the BCL method. During the etching of the NiCr, an abnormality was observed.

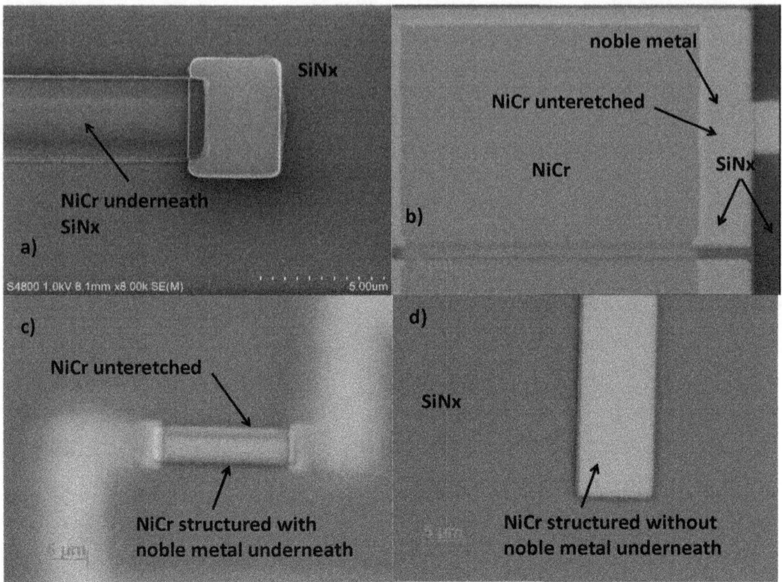

Figure 4.9 Micrographs of the top connection with wet etch (TCW) TFRs underetching by wet etching, a) NiCr structured by wet etching on top of the handover metal, b) NiCr underetched with handover metal underneath the whole NiCr area, c) NiCr underetched beneath SiNx after electroplated gold, d) NiCr not underetched when handover metal is missing.

Fig.4.9 shows the abnormality which occurred during the NiCr wet etching with the BCW method. Subframe b) shows clearly a significant lateral underetching of about 5 µm of NiCr beneath the SiNx etchmask. This is a massive uncontrollable underetching by considering the etch depth of 44 nm to structure the NiCr. This is an acceleration in etch rate by the factor of 100 when comparing the lateral and vertical etching. To understand the root cause for this behavior, subframes c) and d) are helpful. In subframe c) the NiCr is underetched and in d) not. The underetching occurs only when NiCr is connected to the handover metal. Such an effect can be explained by the difference in electrochemical potential between the noble and less noble metals [16]. The connection of the ignoble with a noble metal accelerates the oxidation and leads to a significant faster etch rate of the NiCr [55, 56]. Therefore the BCW approach was not expedient due to the uncontrollable etch rate and therefore leading to a not connected NiCr layer to the handover metal.

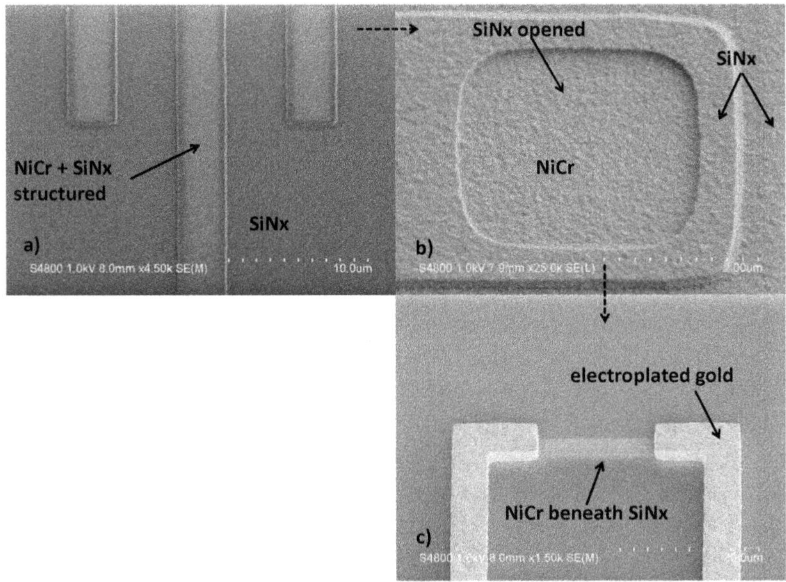

Figure 4.10 SEM micrographs of the TFRs fabrication with the top connection with wet etch (TCW), a) NiCr structured by wet etching, b) SiNx above NiCr opned for connection, c) electroplated gold connected the NiCr from top.

In Fig.4.10, the processing of the TCW method is shown. Here, a handover metal like in BCL and BCW is not needed, since the connection is established from the top to the NiCr. In this method the NiCr and the SiNx is applied on the whole wafer and structured like in the BCW method. In the next step, a 4.4 µm thick AZ nLOF 2035 photo resist is applied to serve as an etch mask. The processing for the opening of the SiNx is done similar to the SiNx structuring for the NiCr wet etch (as in BCW). After the opening, the NiCr is connected by the plating base of the electroplated gold.

Since the BCW method was not successful (NiCr not connected to handover metal), the electrical results only for BCL and TCW are discussed in the following chapter.

4.2.3 Electrical results

Within the TCW method, three different NiCr thicknesses have been evaluated. The common sheet resistance for TFRs in the MMIC technology is 50 $\Omega/\square$ [51]. Nevertheless, 25 $\Omega/\square$ and 10 $\Omega/\square$ sheet resistances have been evaluated.

Table 4.1 Sheet resistance results for $R_s = 10\ \Omega/\square$, $R_s = 25\ \Omega/\square$ and $R_s = 50\ \Omega/\square$. Based on [16].

R_s setpoint [$\Omega/\square$]	thickness [nm]	R_s mean VDP [$\Omega/\square$]	R_s S.D. VDP [$\Omega/\square$]	R_s mean 4P [$\Omega/\square$]	R_s S.D. 4P [$\Omega/\square$]
10	114	9,91	0,19	9,77	0,22
25	44	25,3	0,54	25,24	0,56
50	22	52,2	1,13	51,7	1,1

In Tab. 4.1, the measurements for the three different sheet resistances are shown. In this table, two measurement equipments are compared. The mean VDP measurements have been done on finished wafers with a dedicated on-wafer measurement tool (Karl Süss PA200). Within the VDP measurement, 25 measurement points (shots) are evaluated on each wafer. The measurement layout was designed as in Fig.4.4 in subframe b). The 4P measurement (with Veeco AP-150) was performed on dedicated test substrates which were fully covered with NiCr. Here, a four point measurement is performed 43 times on a 3 inch wafer. As seen, the sheet resistance can be adjusted with the desired value. The measured results with the redundant equipment are in excellent agreement with literature [47, 51]. As described above, 50 $\Omega/\square$ sheet resistance is the common value for MMICs. Nevertheless, 25 $\Omega/\square$ was chosen for the TFR, due to the non-linear dependency between NiCr thickness and resistivity in the range of 1 nm and 15 nm and to avoid yield losses at the lower thickness. The sheet resistance of 10 $\Omega/\square$ with a thickness of 114 nm would perform even better than 25 $\Omega/\square$ but would consume considerably more area for high resistor values [16].

Table 4.2 Experiment matrix for NiCr contacts with the BCL- and TCW method. Based on [16].

pretreatment	#1	#2	#3	#4	#5	#6	#7	#8
TCW method	x	x	x	x	x			
BCL method						x	x	x
O_2-Plasma 1 min	x	x	x	x	x			x
H_2O rinse 10 min	x	x	x	x	x			x
200 °C 10 min	x	x	x	x	x			
Cr-etch18 20 sec		x						
HCl:H_2O 1:10 1 min			x	x	x	x	x	x
Ar sputter time [min]	3	0	0	6	3	1	0	1
ρ_c [$\Omega \cdot cm^2 \times 10^{-8}$]	1890	1790	38.7	26,1	28.8	0.08	0.8	1.11

In Tab. 4.2, the dependency between contact resistance and pretreatment for the BCL and TCW is shown. As described above, no BCW measurements were possible due to the underetching of the NiCr. In the first row, 8 different experiments are depicted. #1 to #5 are fabricated with the TCW and #6 to #8 with the BCL method. For the TCW method, the pretreatment is related to the sequence after SiNx opening and before application of the plating base. In the case of BCL, the pretreatment is related to the sequence before the NiCr is applied on the handover metal. The data points indicate a strong dependency of the pretreatment on the contract resistivity. The sequence of oxygen (O_2) plasma, water (H_2O) rinse and oven bake for the TCW method are necessary for a good plating base adhesion (evaluated in internal experiments at FBH) and were therefore not changed. After the opening of the SiNx for the NiCr contact, the O_2 plasma is needed to remove organic residues on the wafer surface.

As seen, the best contact resistivity from TCW is almost by a factor of 30 higher than the highest value from BCL. The lowest measurable ρ_c with the BCL method (#6) was $8 \times 10^{-10}\ \Omega \cdot cm^2$, which is comparable low and suitable for an MMIC process [51]. The lowest ρ_c with the TCW method (#4) was $2.61\ \times 10^{-7}\ \Omega \cdot cm^2$. With the lowest ρ_c in the TCW method the value is three orders of magnitude higher than in the BCL method. The measurable improvement of the ρ_c by using a diluted hydrochloric acid with and without the in-situ Ar-sputtering directly before the plating base indicates an oxidation of the revealed NiCr opening. This assumption correlates with the knowledge about fast growth of Cr_2O_3 [57, 58]. The presumably fast growing Cr_2O_3 is removable as seen in #3, #4 and #5. In the difference between the samples with and without Ar-sputtering an improvement with the sputtering is measurable. However, with the doubled sputter time in #4 compared with #5, the improvement in the contact resistance is minimal, which indicates a redeposition of SiNx from the surrounding area during the Ar pretreatment.

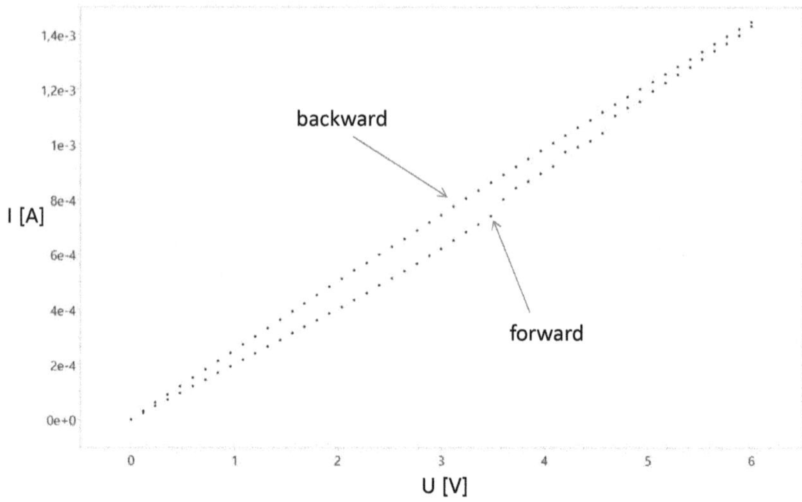

Figure 4.11 IV-curve of #1 NiCr contact from Tab. 4.2.

In Fig.4.11, an IV-curve is shown from the #1 experiment which was described in Tab. 4.2. In this curve, the IV-characteristics of one NiCr contact is shown in both directions. The ideal behavior is in both direction the same IV-path. In Fig.4.11 the backward direction has a slightly higher voltage than in the forward direction. Such characteristic describes a loading (capacitor) effect between the NiCr resistive layer and the TiW platingbase. This indicates the presence of a dielectric material (either grown Cr_2O_3 or redeposited SiNx) between both contact metals.

Nevertheless, with the BCL method and the #6 approach, an ideal process is found for integration in the InP MMIC process. Regarding the amount of necessary lithography layers, the BCL layer is similar to BCW and TCW.

4.2.4 Thermal stress tests and simulation

As described in Chap. 4.2.2, the NiCr TFRs are placed for suitable HF performance on top of the BCB, were the last interconnect metal is located. To fulfill the needs for good thermal conductivity, the NiCr should be placed on the host substrate. Since BCB has a poor thermal conductivity (0.29 W/m K), the TFRs have been stressed to evaluate the maximum applicable power.

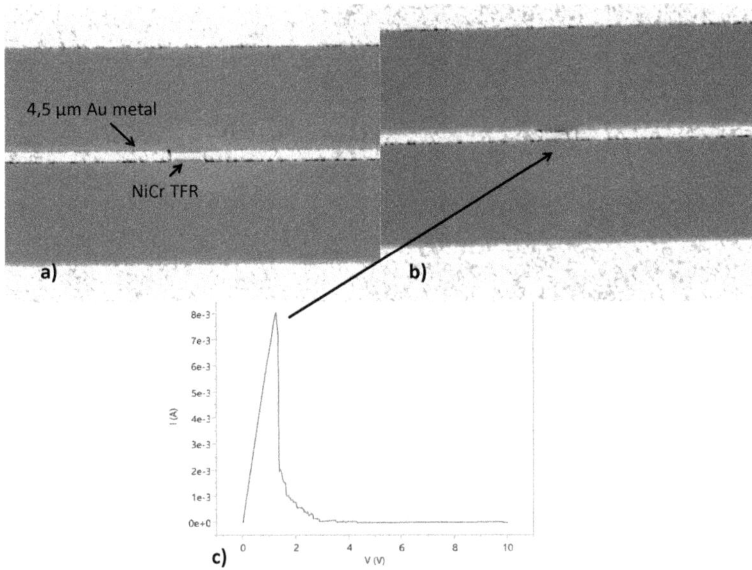

Figure 4.12 Stress test micrographs of a NiCr TFR on top of BCB with a width of 5 µm and a length of 20 µm at a sheet resistance of 25 Ω/□. a) TFR before stress test, b) TFR after stress test, c) IV-graph of the stress test.

In Fig.4.12 an example for a thermal stress test is shown. In this test, the power was swept up to the necessary destructive power. In subframe c) in Fig.4.12, the respective IV-curve for NiCr stress test is shown. A maximum current of 6.5 mA at voltage with a resistance of 100 Ω was measured, which results in a maximum dissipated power for this device of 4 mW. In subframe a), the device is shown before and in subframe b) after the destruction. The failure indicates either damaged SiNx beneath (thermal stress of SiNx or BCB) the NiCr due to the high temperature or a lifted SiNx with NiCr due to the thermal expansion of the resistive metal. Such failure pattern is typical for NiCr thin film resistors [59]. The same experiment was performed with NiCr resistors on aluminum nitride wafers. The intention, was to evaluate the limits of 44 nm NiCr with a good thermal connection.

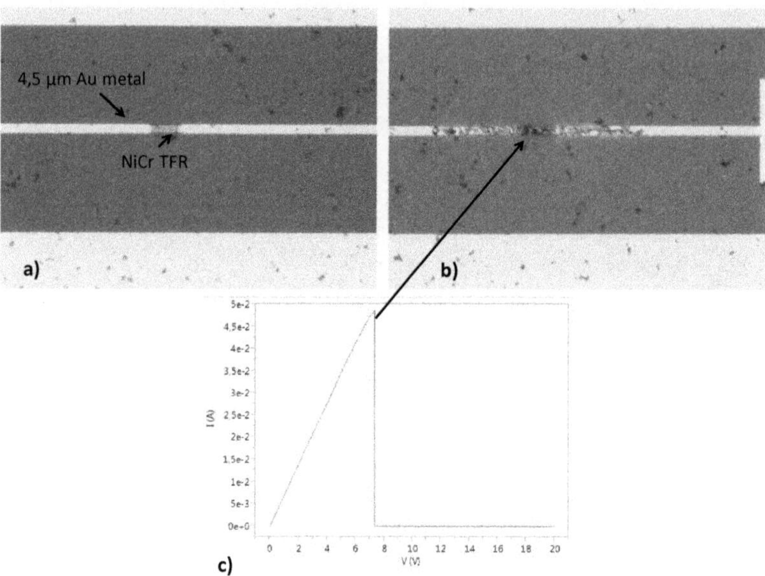

Figure 4.13 Stress test micrographs of a NiCr TFR on top of AlN with a width of 5 µm and a length of 20 µm at a sheet resistance of 25 Ω/□. a) TFR before stress test, b) TFR after stress test, c) IV-graph of the stress test.

In Fig.4.13, the experiment on an AlN wafer, a material with good thermal conductivity and without BCB beneath NiCr, is shown. The resistor layout is similar (same mask layout) as in Fig.4.12. The NiCr TFR with 5 µm width and a length of 20 µm on AlN is able to tolerated by a factor of 50 more current as the TFR on BCB. Here, the drawback of the NiCr TFR placed at the layer of the microstrip line is evident. NiCr TFRs on top of the BCB (at the microstrip line) with the poor thermal conductivity leads to a significantly lower maximum dissipated power than on the AlN host substrate.

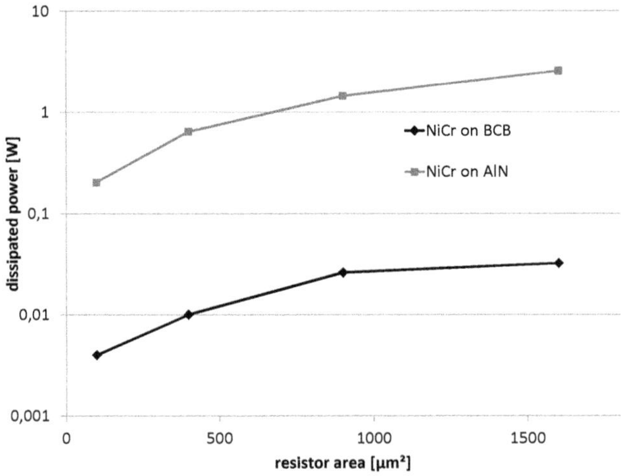

Figure 4.14 Measurement graph for the maximum dissipated power as function of resistor area on AlN and BCB. Based on [16].

In Fig.4.14, the stress test measurements are illustrated for NiCr on AlN and on BCB. The maximum dissipated power is shown as a function of the resistor area. NiCr applied on AlN with a good thermal connection can draw a significant higher power than on BCB. The NiCr thin film resistors on AlN can exhibit power increased by a factor of 50 to 80. The maximum current density for a TFR with 5 µm width and 20 µm length (100 Ω type) is therefore 2.95×10^{10} A/m^2 for BCB and 2.05×10^{11} A/m^2 for AlN, respectively.

Since the same technology in the comparison of NiCr applied on BCB and AlN was used, an electromigration in the case of BCB is unlikely. In the case were NiCr is applied on AlN, an electromigration failure is possible. To evaluate the maximum occurring temperature on the BCB, a thermal simulation was performed.

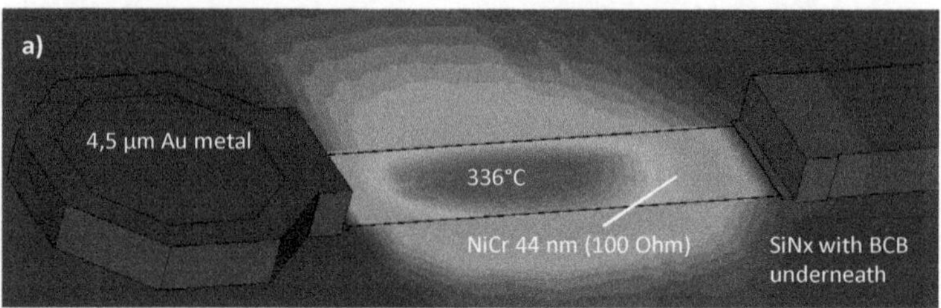

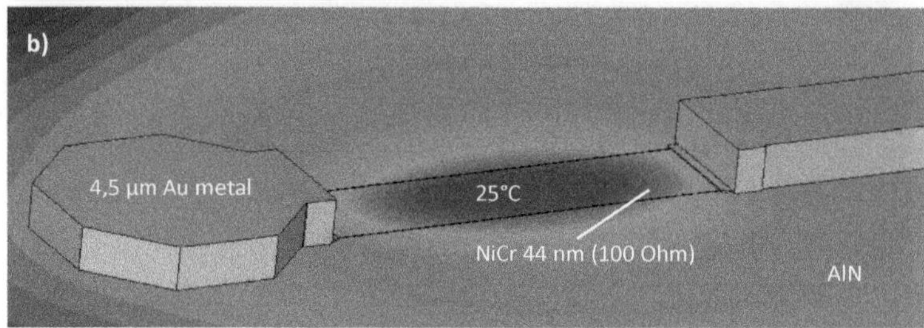

Figure 4.15 Thermal simulation with CST Studio Suite 2017 of a NiCr TFR with 100 Ω with an area of 10 μm × 40 μm and 0.01 W dissipated power. a) NiCr on BCB (as in Fig.4.12), b) NiCr on AlN (as in Fig.4.12). Based on [16].

In Fig.4.15, the thermal simulation of NiCr on BCB in subframe a) and on AlN in subframe b) is shown, respectively. For simulations, the thermal conductivity of 0.29 W/m K for BCB, 315 W/m K for Au, 15 W/m K for NiCr and 160 W/m K for AlN was chosen [17, 60–62]. The base plate temperature was set to 20 °C. The simulation confirms the stress measurements, where NiCr applied on top of BCB is heating up to a critical temperature for BCB. The simulated temperature of NiCr on BCB exceeds a value of 336 °C, whereby on AlN with the same thermal stress and setup the NiCr stays at 25 °C.
To evaluate the validity of the thermal simulation, a temperature coefficient (TCR) measurement was done. With such TCR, which is material specific, the resistance as a function of temperature can be calculated and vice versa [47]. In Eq. 4.10, the calculation of the TCR is shown.

$$\alpha = \frac{1}{\rho(T_1)} \left(\frac{\rho(T_2 - \rho(T_1))}{T_2 - T_1} \right). \tag{4.10}$$

The resistive element must be applied on a thermal conduction wafer or substrate and placed on a thermal chuck for the measurement. For electrical measurement, the temperature is

swept from room temperature up to a sufficient temperature to measure a difference in resistance. In this case, the temperature was swept from 20 °C up to 95 °C with NiCr applied on AlN as in Fig 4.13. The resulting TCR for NiCr is 162 ppm/K. The TCR is in agreement with typical literature values [63, 64, 21, 47, 51]. With the TCR and the measurement points from Fig.4.12, a temperature of 298 °C, for 10 mW power dissipation with NiCr on BCB was calculated [16]. With this calculation, the simulated temperature as shown in Fig.4.15 is reasonable.
In conclusion, the thermal aspects need to be taken into account when implementing the NiCr on BCB into the InP MMIC process. One possibility is to split one 50 Ω resistor in two parallel 100 Ω resistors to mitigate the risk of thermal damages.

4.3 Implementation into the InP HBT process

Since the NiCr TFRs were successfully developed as described in Chap. 4.2.2, the implementation of the TFRs into the InP process was necessary. Therefore the layout and the design rules have been generated.

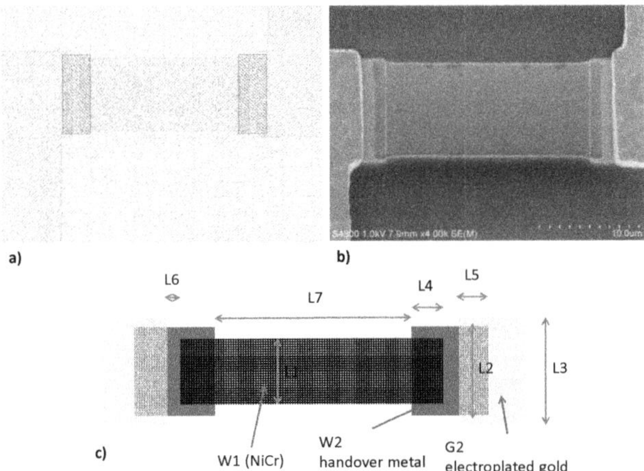

Figure 4.16 NiCr TFR layout and design rules for the utilization in the InP MMIC process. a) layout design of a typical NiCr TFR, b) SEM micrograph of a finished TFR, c) design rules for the TFR.

In Fig.4.16, the NiCr resistor layout is shown. In subframe a) and b), the top view of a NiCr resistor is shown in the layout and as SEM micrograph, respectively. In subframe c), the design rules are illustrated. The design rules have been chosen by considering the alignment accuracy of the lithography tool (Nikon i12) and process variation.

Table 4.3 Design rules for the NiCr TFR in the InP MMIC technology.

Design rule	min. distance [μm]	Description
L1	5	min. NiCr width (W1 layer)
L2	6	min. handover pad width (W2 layer)
L3	7	min. G2 contact width (G2 layer)
L4	1.5	min. NiCr overlap over W2
L5	1.5	min. W2 overlap over G2
L6	1	min. space between NiCr and G2
L7	5	min. NiCr length (W1 layer)

In Tab. 4.3, the design rule values are described. The effective area of the resistive element is the length L7 times the width L1. Another important factors for suitable HF performance, the widths L1, L2 and L3, have been layouted graduated but kept at 0.5 µm increase for each side. With a significantly wider difference between L1, L2 and L3, an unwanted behavior (HF stub) can occur and degrade the HF performance. The design rules and area calculation have been implemented in the design tool (Cadence).

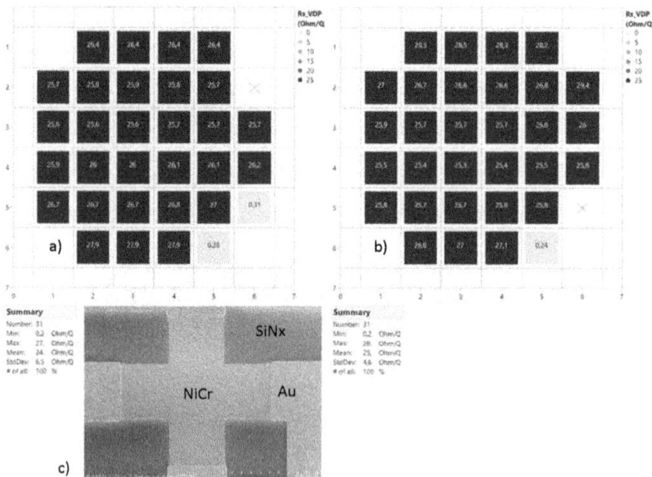

Figure 4.17 Sheet resistance distribution on a finished InP MMIC process. a) first MMIC run, b) second MMIC run, c) SEM micrograph of the VDP measurement structure.

In Fig.4.17, the sheet resistance distribution of two finished InP MMIC runs are shown. Those two runs have been finished sequentially with one year apart. The distribution of the NiCr sheet resistance is an indicator for the thickness homogeneity across the 3 inch wafer. The resulting sheet resistances for full wafer maps are 24.6 $\Omega/\square$ (S.D. 6.52 $\Omega/\square$) and 25.6 $\Omega/\square$ (S.D. 4.84 $\Omega/\square$) for subframe a) and b), respectively.

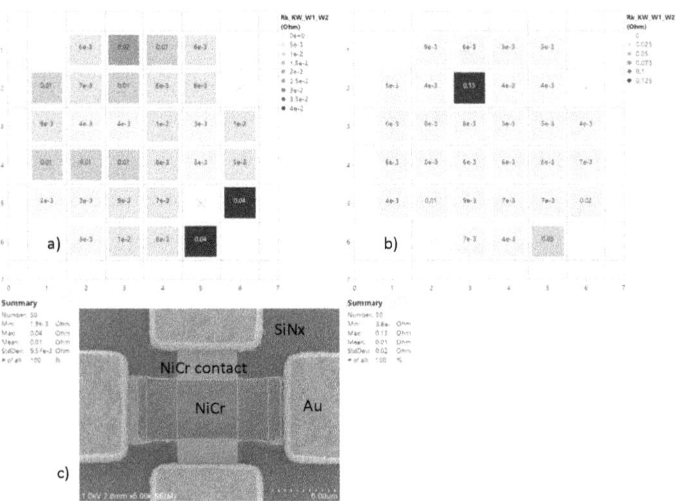

Figure 4.18 NiCr TFR contact resistance distribution between NiCr and contact metal in a finished InP MMIC process (one wafer). a) first MMIC run, b) second MMIC run, c) SEM micrograph of the VDP measurement structure.

Fig.4.18 shows the distribution of the NiCr contact resistance to the handover metal. As in Fig.4.17, both wafer maps were used on two wafers from two different (chronologically separated) runs. The resulting contact resistance was 0.01 Ω (S.D. 0.009 Ω) and 0.01 Ω (S.D. 0.02 Ω) for subframe a) and b), respectively. With the effective area of 5 µm x 5 µm and the contact resistance of 0.01 Ω, a contact resistivity of $2.5 \times 10^{-9}\ \Omega \cdot cm^2$ can be calculated. Such value is close to the contact resistivity of $8 \times 10^{-10}\ \Omega \cdot cm^2$, which was extracted by the TLM method in Chap. 4.2.3. With this measurement, a suitable contact resistance of the NiCr TFRs was confirmed on actual wafer runs.

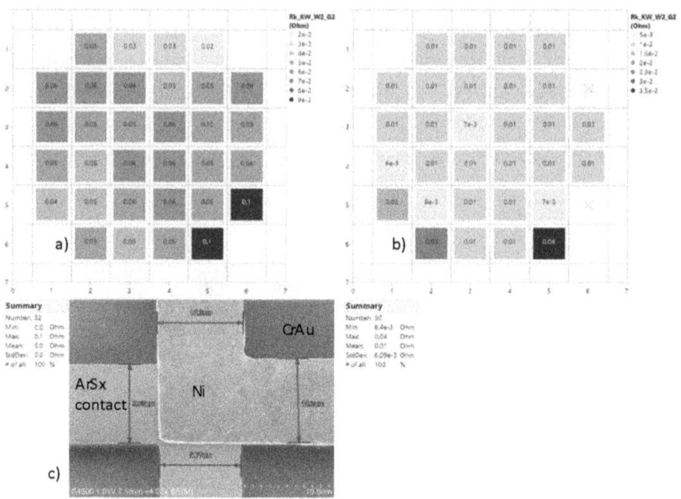

Figure 4.19 NiCr TFR contact resistance distribution between contact and interconnection metal in a finished InP MMIC process (one wafer). a) first MMIC run, b) second MMIC run, c) SEM micrograph of the VDP measurement structure.

In Fig.4.19, the contact of the handover metal to the interconnection metal is shown. The two wafer maps have been measured on two separate wafer runs, as shown in Fig.4.17 and Fig.4.18. The resulting contact resistance is 0.06 Ω (S.D. 0.01 Ω) and 0.01 Ω (S.D. 0.006 Ω) for subframe a) and b), respectively. With 0.01 Ω and an effective area of 10 µm x 10 µm a contact resistivity of $1 \times 10^{-8}\ \Omega \cdot cm^2$ can be calculated. For both, the contact NiCr to handover metal and handover metal to interconnect metal the resistance suits perfectly in a MMIC. In the case of a contact resistance with several orders of magnitude higher than $1.9 \times 10^{-5}\ \Omega \cdot cm^2$ (top connection fabricated by wet etch), the NiCr TFRs would not work properly in the application. With the BCL fabrication method (bottom contact and fabricated by lift-off as described in Chap. 4.2.2) the NiCr resistors are applicable in the InP MMICs.

4.3.1 High frequency results

The NiCr resistors have been also implemented for high frequency measurements and in InP MMIC circuits. As mentioned in Chap. 4.1, thin film resistors can be utilized for several purposes in the high frequency applications.

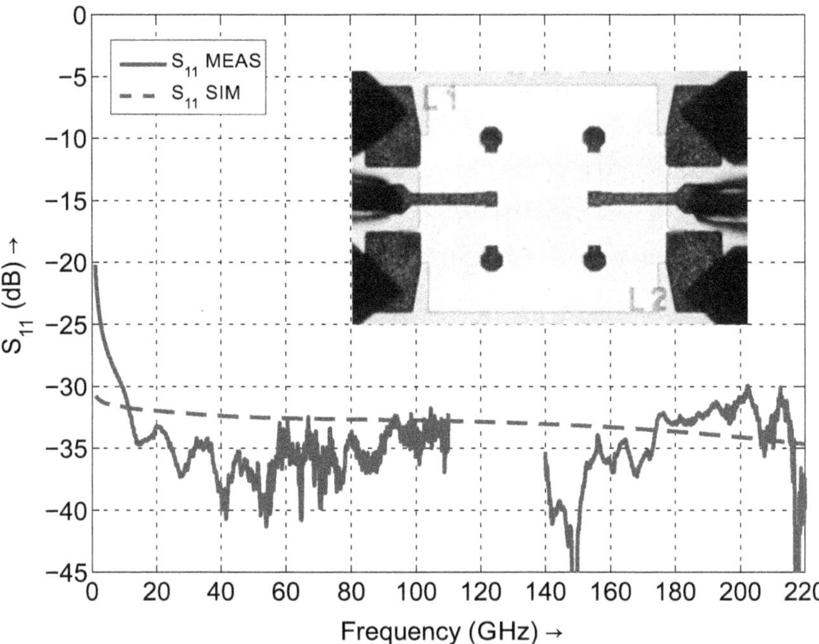

Figure 4.20 Reflection coefficient S_{11} of the 50 Ω NiCr termination versus frequency from on-wafer measurement (solid) and 3-D EM simulation by Dr. M. Hrobak (dashed) [16].

The NiCr TFRs have been implemented in several actual MMIC process runs, as mentioned in Chap. 5.4. In Fig.4.20, the measurement of a 50 Ω NiCr termination is shown. The measurement was performed on-wafer with scattering parameter from 10 MHz to 220 GHz after multiline thru-reflect-line (mTRL) calibration [65]. The measurements are in good agreement with 3-D electromagnetic (EM) simulations. Since the NiCr sheet thickness was processed with 44 nm, the sheet was modelled as an impedance boundary with zero thickness. In Fig.4.20, the reflection coefficient S_{11} of a measured 50 Ω termination is compared with simulation. The termination is realized with two 100 Ω resistors. The measured and simulated return loss values are greater than 30 dB in the frequency range of 5 to 220 GHz. Such return losses are therefore suitable for an on-wafer LOAD calibration standard [16]. Such broadband terminations as shown in Fig. 4.20 are among others part of traveling wave amplifiers (TWA) [66] and directional couplers [67].

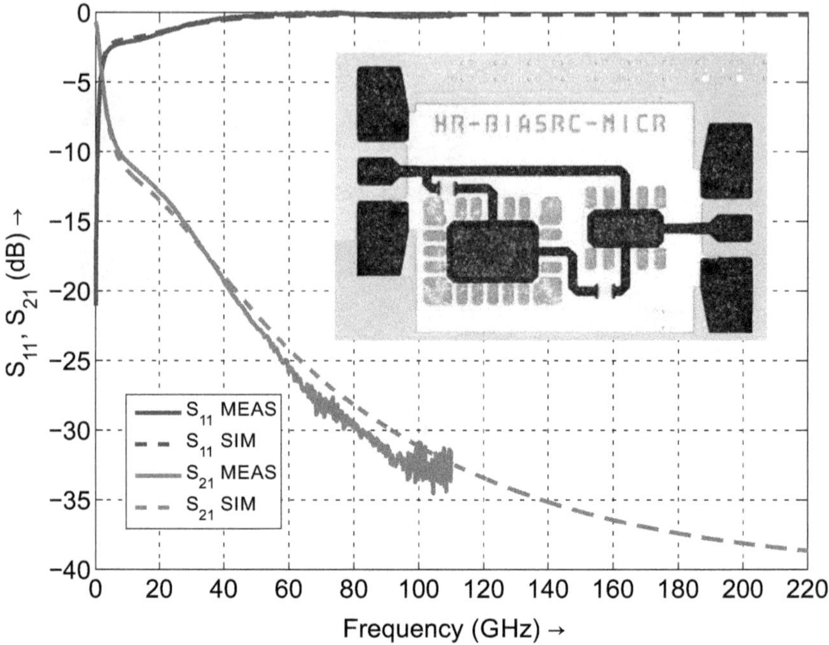

Figure 4.21 Reflection S_{11} and transmission S_{21} coefficients of the mmW choke versus frequency from on-wafer measurement (solid) and 3-D EM simulation by Dr. M. Hrobak (dashed) [16].

In Fig.4.21, the measured and simulated reflection S_{11} and transmission S_{21} coefficients of a mmW choke are compared. Such choke is used e.g. in a biasing network of a W-band five-stage low noise amplifier (LNA) [68]. As stated in Chap. 4.1, NiCr resistors are also used for Wilkinson power dividers. In the case of this dissertation, the NiCr TFRs were used to suppress differential mode excitation and to achieve isolation of Wilkinson power dividers (100 Ω) in InP power amplifiers. The InP power amplifiers can suffer from thermal runaway and current hogging, which can be mitigated or even avoided by utilization of NiCr base or emitter ballasting resistors [69].

Feedback resistors can be used in a stability network as part of the DHBT base and collector biasing within transimpedance amplifiers (TIA) [66]. With the high sheet resistance and sufficient current carrying capability, the NiCr TFRs are indispensable in compact MMIC designs for high frequency applications. The TFRs have been used in traveling wave amplifier, as shown in Fig. 4.22.

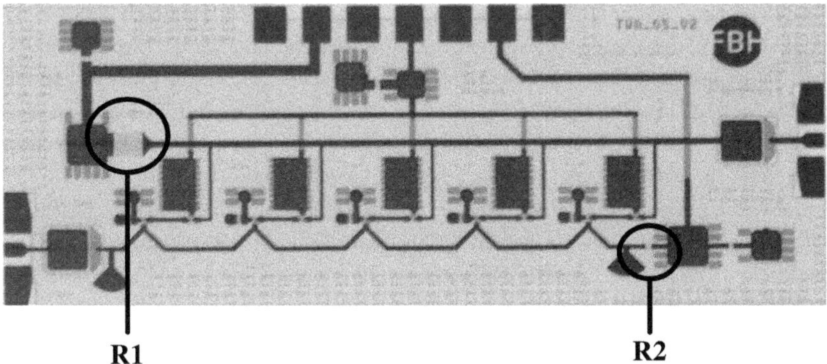

Figure 4.22 95 GHz traveling wave amplifier using NiCr resistors for RF termination and DC bias. Based on [70].

The traveling wave amplifier shown in Fig.4.22, consists of five stages, where in each stage two transistors are connected as cascode configuration. To achieve a suitable matching at the in- and output, two terminations with R1 = 35 Ω and R2 = 29 Ω were implemented. Both terminations serve as resistive paths for DC and HF functionality, which also demonstrate the wide broadband behavior of the TFRs. In HF measurements the TWA shown in Fig.4.22, exhibited a small signal bandwidth of 95 GHz and 12 dB forward gain from input to output [70].

With the HF and circuit measurements, the NiCr TFRs are now tested and verified as a suitable and essential element in InP MMICs.

4.4 Chapter 4 summary

In Chap. 4, the development and implementation of NiCr TFRs is shown. The motivation for a resistor element with high resistivity is demonstrated and basic fundamentals of TFRs in MMICs introduced. The difficulties and challenges during the process development have been shown. The NiCr TFR contacts from top and structured by wet etch, resulted in high NiCr to metal resistance ($2.61 \times 10^{-7}\ \Omega \cdot cm^2$). I-V curves (top contact) show a capacitive behavior, which indicates the presence of a dielectric (Cr_2O_3) or redeposited material (SiNx). The lowest NiCr to metal contact resistance was achieved by connection from the bottom and structuring by lift-off ($8 \times 10^{-10}\ \Omega \cdot cm^2$). NiCr connected from the bottom and structured by wet, revealed a significant underetching of NiCr (underneath SiNx) due to the occured electrochemical potential between the noble and less noble metals (NiCr and handover metal).

Electrical stress measurements indicates a high thermal resistance of the NiCr TFRs on top of the BCB. With typical power applied, the NiCr resistors (simulations) exceeded 300 °C. NiCr was implemented as a passive element into the actual InP MMIC process. The DC results from two separate MMIC runs, showed a robust and suitable distribution of the sheet resistance, NiCr to handover metal and handover metal to interconnect contact resistances. Therefore, the NiCr resistors were successfully integrated into the MMIC process. HF measurements of passive circuit elements are in good agreement with simulations. Successful circuit measurements with implemented NiCr resistors, demonstrated the last step of the TFR integration. Concluding, the NiCr TFRs are now implemented into the design kit of the InP MMICs at the FBH and are indispensable for broadband and complex circuit topologies.

Chapter 5

Through silicon vias in a transferred-substrate process

5.1 Motivation

As already mentioned in chap. 1, this chapter provides further insight into the technology of through silicon vias (TSVs), including the motivation for the work on TSVs in the framework of this dissertation. Simulations, process development and finally the implementation of TSVs into the TS process are discussed.

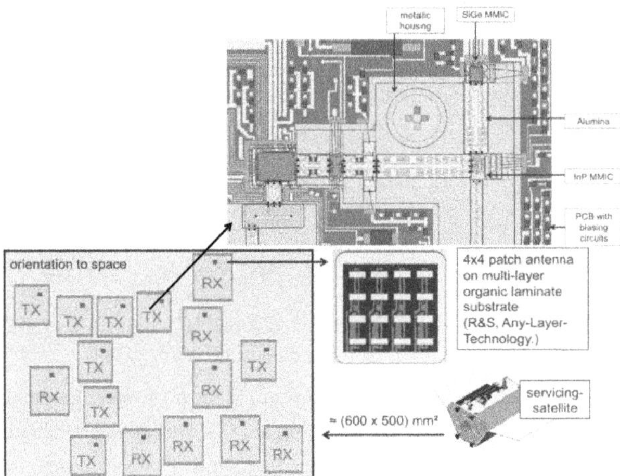

Figure 5.1 Illustration of the radar module for a DLR project (50RA 1327). Based on [71].

After chip fabrication, the integrated circuits are either packaged into a housing or mounted in a system. In this case, a demonstrator for a DLR project had to be assembled. As seen in Fig.5.1, the demonstrator is designed for a service satellite with a millimeter wave positioning system. The main purpose is to perform a de-orbiting maneuver for satellites at the end of life. As deliverable for the demonstrator, tracking with 1000 m resolution using the millimeter wave radar was required. The benefit of such millimeter wave positioning system, is the independency from sun or other light sources. The module consists of numerous transmitting (Tx) and receiving (Rx) parts. Each Tx and Rx part has a 4x4 patch antenna. In each Tx and Rx part many electronic components (power amplifier, mixer, oscillator, up- and down converter etc.) are included. Many electronic components are commercially available, except for a mixer in this frequency range. Such a mixer is shown as InP MMIC in Fig.5.1.
For the demonstrator, the wire bonding assembly approach was chosen, due to the compatibility to the commercially acquired ICs. While the wire bonding of small MMIC pads (e.g. 40x80 μm) is an own topic and is not part of this thesis, the shielding of the MMICs has a huge impact on the performance. Simulations with MMICs in the mounted system indicated massive parasitic modes [71].

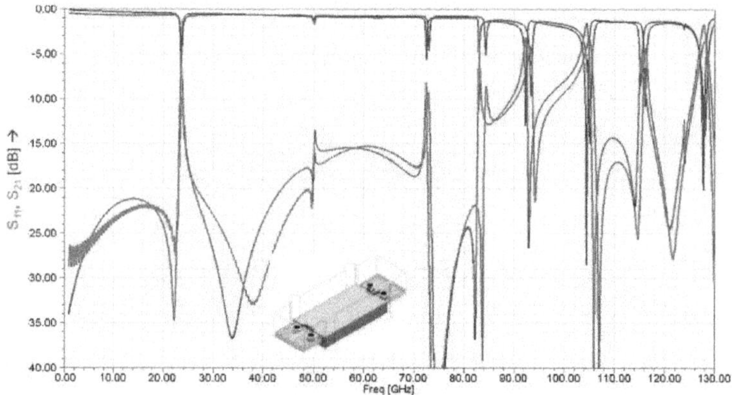

Figure 5.2 High frequency simulation by Dr. M. Hrobak (CST and Ansys HFSS) of an IC (simple transmission line) mounted in alumina with bond wires [71].

As depicted in Fig.5.2, the parallel plate modes occur at different wave lengths. As seen, the input reflection coefficient (S_{11}) and the forward gain (S_{21}) are both showing the unwanted parasitic behavior. The parasitic modes depending on the substrate material (dielectric constant), the substrate thickness and the chip size. Different approaches are possible to avoid or even suppress these unwanted modes.

The easiest way would be a grounded metal plate beneath the MMIC to short circuit the modes. Although this would be a cheap and an easy way to suppress the parasitic modes, it would also degrade all series active and passive elements in the MMIC. Such elements are not shielded with the ground plane (GD) to minimize the capacitive impedance part to the ground. This affects in this case the common base transistors, the series capacitors and the collector lead of the common emitter transistors.
Another possibility is to thin down the substrate (less then 10 μm) with grinding and chemical mechanical polishing (CMP) [22]. This approach requires equipment to process and handle such thin substrates.
The third way is the implementation of grounded TSVs. Those TSVs act like a Faraday cage and short circuit the parasitic modes without harming the high frequency behavior of the active and passive elements [72].

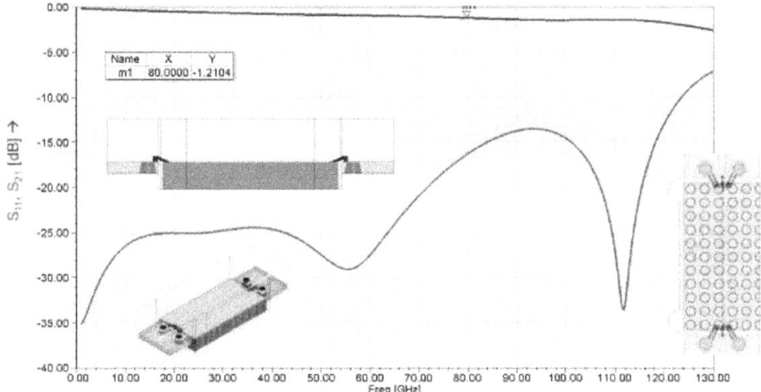

Figure 5.3 High frequency simulation by Dr. M. Hrobak of an IC (simple transmission line) mounted in alumina with bond wires and TSVs implemented [71].

As seen in Fig.5.3, the implementation of grounded TSVs completey suppresses the parasitic modes. In this simulation, TSVs with a conservative diameter of 135 μm were assumed. These results clearly show the potential of implementation of such TSVs and acted as the main motivation for the research in this field. Since the radar module would not work without the implemented TSVs in the InP MMICs, implementation of TSVs into the InP DHBT process was mandatory.

5.2 Fabrication methods of through silicon vias

Through substrate connections were developed for 3D integration in the silicon technology. The purpose of the TSVs described here, is the suppression of parasitic modes. In general, three fabrication methods are available to create TSVs [73]:

- wet etching silicon with potassium hydroxide (KOH) to create an angle of 54.7°
- dry etching vias with the BOSCH- or cyro-process
- laser drilling the substrate interconnects

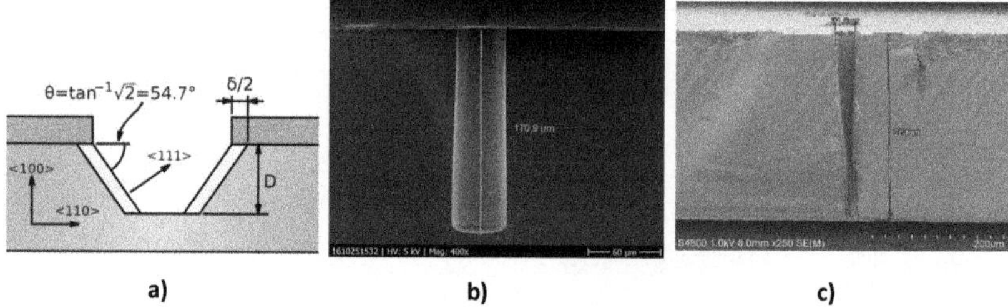

Figure 5.4 Illustration of three different methods to create through silicon vias, a) via wet etch with KOH at an angle of 54.7° based on [74], b) dry etched vias, c) laser drilled vias.

As seen in Fig.5.4, there are three possible methods to create TSVs. It is possible to fabricate vias with a big aspect ratio by dry etching and laser drilling (subframe b) and c) in Fig.5.4). With the wet etching method, the aspect ration is drastically limited by the etch angle of 54.7°.

Due to missing equipment for a BOSCH- and cyro-process at the time of this thesis, the dry etch method was not possible. The wet etch method was available, but due to the angle of 54.7°, the resulting via would consume a big part of the MMIC area. The demonstrator boards had slots of 5 mil (etch depth of 127 μm), which would result in a via diameter of 358.72 μm at an etch angle of 54.7°. Wet etch did not pose a viable option and was disregarded for further investigations.

Table 5.1 TSV fabrication methods in comparison. Based on [73].

	wet etching	dry etching	laser drilling
fabrication speed	1...11 μm/min	up to 50 μm/min	2400 vias/sec
align accuracy	mask defined	mask defined	optical align (few μm)
aspect ratio	1:1...60	1:80	1:7
hole quality	excellent	good (scallops in BOSCH process)	very good
critical dimension	sub μm	sub μm	10 μm

As shown in Tab. 5.1, the three fabrication methods can compete between each other at low or medium requirements. The most common method is the dry etching either with BOSCH- or cyro-processes due to the excellent control of etch rate, via angle, sidewall quality and the highest aspect ratios. Laser drilling is less common due to the lower alignment accuracy which is important in the very-large-scale integration (VLSI). The wet etch method is an uncommon method to fabricate TSVs. Recently published results show promising utilization of the metal-assisted chemical etching (MaCE) technique [75].
Because of the missing deep reactive ion etching (DRIE) equipment and the aspect ration limitation of KOH etched TSVs, laser drilling was selected as method of choice to fabricate the through substrate vias.

5.3 Process development

5.3.1 Methods and Tools

At the FBH, a commercial laser micro machining system (ILS 500S-Air from InnoLas GmbH) was used for the following investigations. The system has two possible laser sources. The laser source for TSV drilling was a high-power Q-switched solid-state UV laser (Coherent AVIA 355-4500). The solid-state laser was diode-pumped at 4.5 W which enables nanosecond pulses at 355 nm with a pulse repetition frequency up to 100 kHz. The computerized numerical control (CNC) of the systems allows a stage control of 200x200 mm^2 in X and Y. The precision control of the laser is performed by a galvo scanner with a field size of 10x10 mm^2. With optical alignment a precision of ± 1 μm is possible. With the telecentric F-theta objective ($f = 56$mm) a laser spot of 10 μm to 20 μm in diameter is feasible [76]. Further used equipment:

- scanning electron microscopy (Hitachi S4800)
- plasma etcher SI 500 from Sentech Instruments GmbH
- i-line step-and-repeat lithography tool (Nikon i12)
- dedicated BCB curing oven (YES-PB6-2PCP)
- wafer bond aligner EVGroup EVG 420
- wafer bonding system EVG®501
- mechanical polisher PM5 from Logitech
- manual spinner for BCB planarization from Süss MicroTec
- electroplating with a commercial system from Mikro- und Oberflächen Technik GmbH
- debris cleaning with a DCS 1440 from Disco
- white light interferometer (optical profiler) from Zygo
- mechanical profiler P10 and P16 from KLA
- metal evaporator for landing pad and alignment marks LH560UHV from Leybold
- DC measurement tool PA200 for (dedicated for process control measurements) from Karl Sss PA200

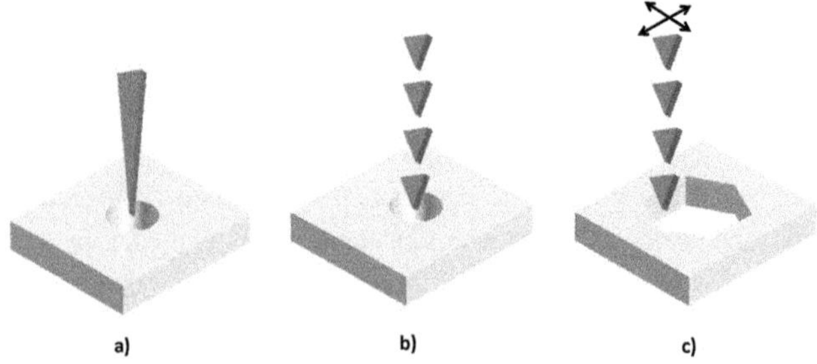

Figure 5.5 Illustration of three methods to drill holes with a laser, a) single beam on one point, b) percussion (single pulses), c) trepanning (laser pulses in any shape).

As seen in Fig.5.5, three options to drill holes with a laser are available. The easiest way to use a single beam and maintain one position for drilling. This has the drawback of limited via shape definition, since this is given by the optics of the system (with this system 10 μm to 20 μm). The same drawback is present at the percussion method in subframe b) in Fig.5.5 but with the advantage of pulsing the laser. By pulsing a laser, less heat is generated inside the laser which gives the opportunity of using higher output power at the same thermal stress of the laser. The third method (trepanning) was used to generate the shapes for the following investigation. With this method, short laser pulses (20 kHz) are redirected by the galvo scanner in X and Y on the substrate.

A single completion of the drilled via area is not sufficient to drill a hole in silicon of more than 100 μm. The power of the laser was set to 0.45 W with a feed speed of 1200 mm/ min.

5.3.2 Mimic process

As already mentioned in Chap. 2, the InP transferred-substrate process has the benefit of using any 3 inch substrate as a carrier. TSVs can be fabricated with the via first, middle and last method. During the InP process at the FBH, a wafer bond in the middle of the fabrication is necessary to transfer the InP substrate to a carrier substrate. This enables the opportunity to utilize the via first method. Via first has a significant benefit in the complex InP TS process of minimizing the risk to damage the finished MMICs at the end of the process. Here the prefabricated carrier substrates with the TSVs can be processed in parallel with the actual process and the most suitable wafer in terms of wafer bow, via yield, quality, etc, can be chosen for wafer bonding to the InP MMICs. First experiments have been performed to see the necessary amount of repetitions to drill circa 130 μm. Therefore, a silicon wafer was processed with vias (diameter of 1 mm).

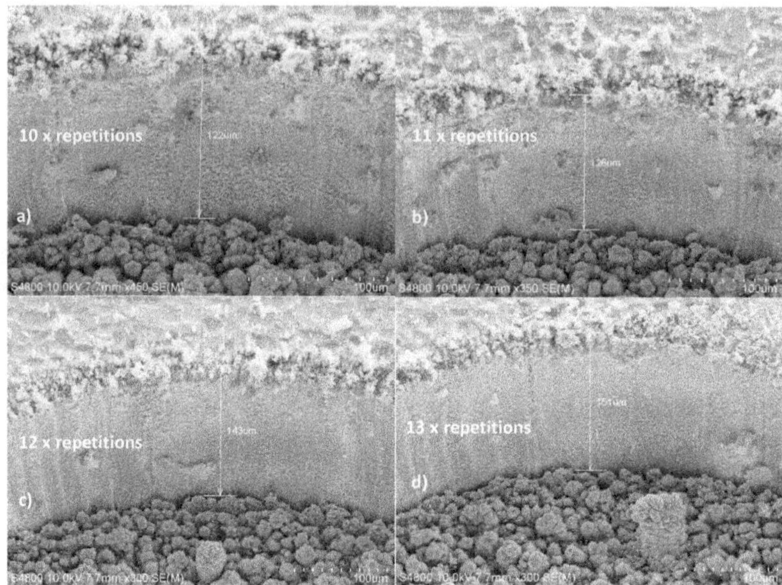

Figure 5.6 SEM micrographs (30 ° tilted) of drilled vias on a 3 inch silicon test wafer, a) 10 repetitions, b) 11 repetitions, c) repetitions, d) repetitions.

As depicted in Fig.5.6, each repetition drills circa 120 μm deep. For the radar module a 127 μm (5 mil) thick substrate is needed. Therefore, it was decided to use 11 repetitions, since the result showed a drill depth of more then 126 μm. With the experiment from Fig.5.6, the repetition was set to 11. In the next step of the investigation a mimic experiment was performed to see the feasibility of the laser drilled TSVs. The visible dirt in and outside of the vias, are residues (debris) which occur during the laser processing. A vacuum cleaning pipe is attached near the laser, that removes the major part of the residues during processing. The remaining residues (mostly oxidized silicon) can be removed afterwards. The process of via cleaning after the laser processing is shown in this chapter. With the evaluated amount of repetitions for laser drilling, the process development was started.

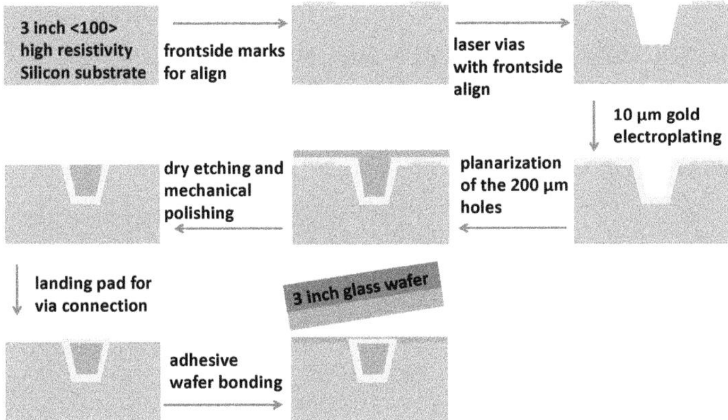

Figure 5.7 Illustration of the mimic experiment for TSVs investigation. Based on [77].

As seen in Fig.5.7, 3 inch $< 100 >$ substrates were used for the experiments. To evaluate the alignment accuracy, marks were patterned by a lift-off process on the front side of the substrate (in the actual process the marks are on the backside as seen in Chap. 5.4). In the next step, the laser system was aligned to the front side of the wafer using the alignment marks. Due to the fact, that the laser optics have an effective focus depth of 70 μm, the final angle of the TSVs have an angle of less then 90°. The limited focus depth and reflecting laser beams from the sidewall to the via bottom of the via, lead to a slanted angle of the TSV where the bottom is narrower than the via top. This is necessary for the following process steps (metalization and planarization). In the next step, the whole front side is metalized (including the vias). After the metalization, the vias have to be planarized. A void free TSV is mandatory for the adhesive wafer bonding. Furthermore, the protruding BCB is removed by plasma etching. The top electroplated gold is removed with mechanical polishing. In the last step before the wafer bonding, a landing metal pad is fabricated on top of the TSV for connection between the TSV and the via from the ground (V1X) reaching from the InP process. In the case of the mimic experiment, the landing pad is used for electrical measurements of the TSVs. At the end of the TSV process, the finished host wafer (with TSVs) is bonded to the active InP wafer or in the mimic process to a glass wafer to evaluate the post-bond quality.

In the mimic process, two layers from an existing layout were used. Those layers are the K1 (see Fig.2.10 in Chap. 2.2) and C (see Fig.2.14 in Chap. 2.2). The K1 layer was used for alignment marks and the C1 with many different shapes and sizes (capacitors) for the TSVs. After applying the marks, an alignment test was performed.

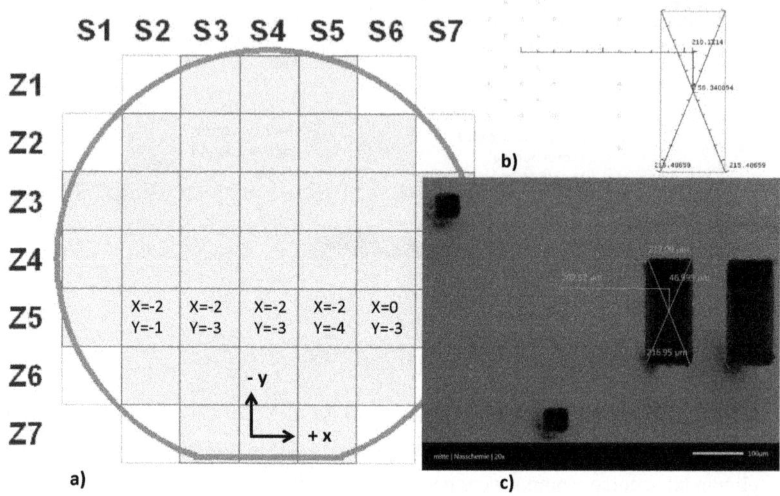

Figure 5.8 Align test of the laser system. a) measurement data of 5 shots with x and y misalignment, b) layout design of the structure overlay, c) micrograph of the alignment measurement.

The alignment accuracy is sufficient as seen in Fig.5.8. The misalignment is within a few micrometer, which can be compensated with an appropriate diameter of the TSV (50 μm to 60 μm). Since the V1X via, which connects the ground to the landing pad of the TSV is circa 15 μm, a misalignment sum of 35 μm to 45 μm is acceptable for wafer bond and TSV alignment. For alignment, points with the most possible distance were choosen, for the best possible alignment quality. As seen, 5 points on the wafer have been measured to evaluate the misalignment (0 μm to 4 μm). In the next step, methods have been evaluated to clean the debris after laser drilling the TSVs. During the drilling, a mixture of silicon and silicon oxide is deposited on top and inside the via.

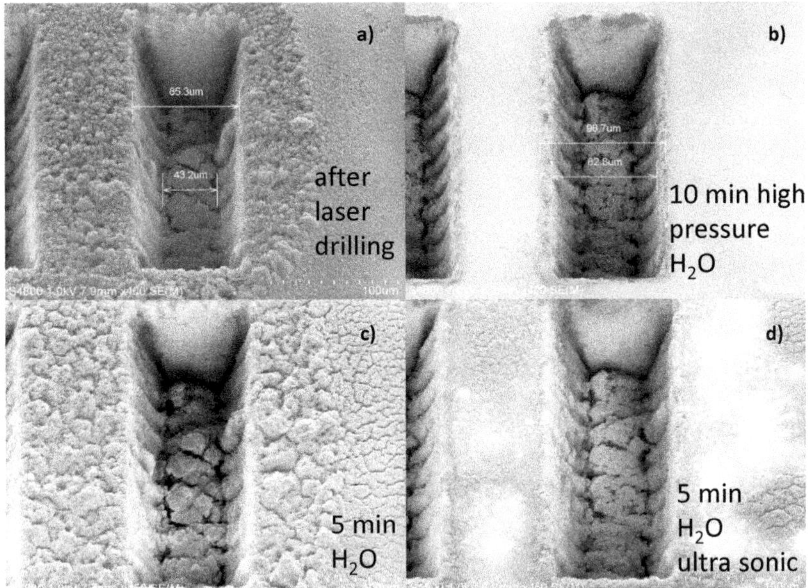

Figure 5.9 SEM micrographs (30° tilted) of the evaluation for cleaning methods after laser drilling. a) after drilling, b) 10 min H_2O cleaning with high pressure (20 MPa), c) 5 min in a H_2O bath with agitation, d) 5 min H_2O bath with ultra sonic.

While hydrofluoric acid is capable of cleaning the debris after laser drilling, a less aggressive solution was desired. Since the alignment marks always contain a first layer of titanium for adhesion, the danger of loosing marks on the backside (which are needed for wafer bonding) is not acceptable. Therefore, methods with H_2O were evaluated as seen in Fig.5.9. The best results have been achieved with a commercial high pressure cleaning tool. Here, a water pressure of 20 MPa was applied on the silicon wafer to clean the TSVs. The ultra sonic cleaning was also helpful, but not sufficient, as seen in the subframe d). Simple water rinse had no visible impact of the debris. In the next step, the aspect ratio capability of the laser system was evaluated. Therefore, many silicon pieces (1x1 mm) have been drilled with various diameter and scribed for a cross sectional view.

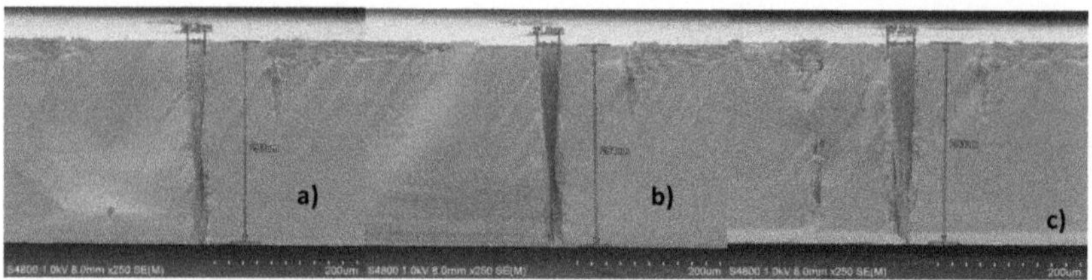

Figure 5.10 SEM cross section micrographs of the aspect ratio evaluation. a) 20 μm via, b) 25 μm via and c) 30 μm via.

In Fig.5.10, the capability of the laser system is shown. A diameter of circa 20 μm is possible with a clear depth of more then 150 μm. This is perfectly suitable for the TSVs in the InP process. The deeper rift beneath 150 μm in the smaller via is due to the reflecting laser beam at the via wall. With bigger vias (50 μm to 60 μm), which are needed for the InP TSV process, the drill depth correlates perfectly with the demands for the TSV process.

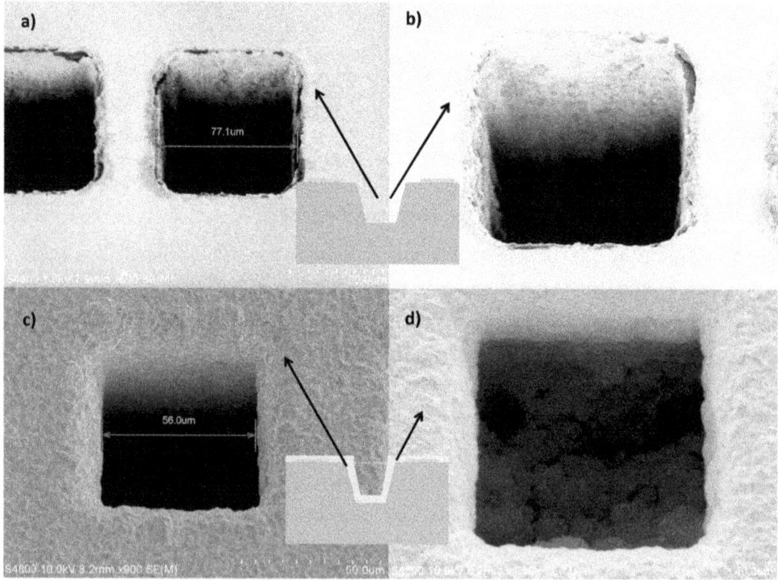

Figure 5.11 SEM micrographs (30° tilted) of laser drilled and electroplated vias. a) and b) after laser drilling and debris cleaning, c) and d) after electroplating the whole wafer with 10 μm gold.

Fig.5.11 depicts the sample after laser drilling and electroplating. The drilling was performed as pointed out in Fig.5.6 with 11 repetitions and the high pressure water cleaning. For electroplating, a seed layer of 100 nm titanium and 200 nm gold was deposited. The seed layer was applied by DC-sputtering to achieve a conformal deposition of the metal. The electroplating was performed by pulse plating with an on- and off-time of 10 ms and 40 ms, respectively. The bath temperature was set to 40 °C and an electrical charge of 450 As at a pulse current (peak current during the 10 ms) of 0.36 A. Unlike the usual copper electroplating of TSVs, in this process, copper plating is incompatible. While it is possible to dope InP single crystals with copper to change the electrical and optical properties of InP, in this case an impurity would degrade the devices [78]. Additionally, a copper plating bath is not desired at the FBH, due to its ability to change of material parameters of the compound semiconductors.

The planarization of the circa 200 μm deep and 50 μm wide via required more pretreatment than the typical planarization with BCB. In the first planarization tests the limits of the outstanding planarization capability of BCB was experienced. After the electroplating, an 80 nm thick silicon nitride layer was deposited. This is necessary due to the poor adhesion of BCB on noble metals like gold [79].

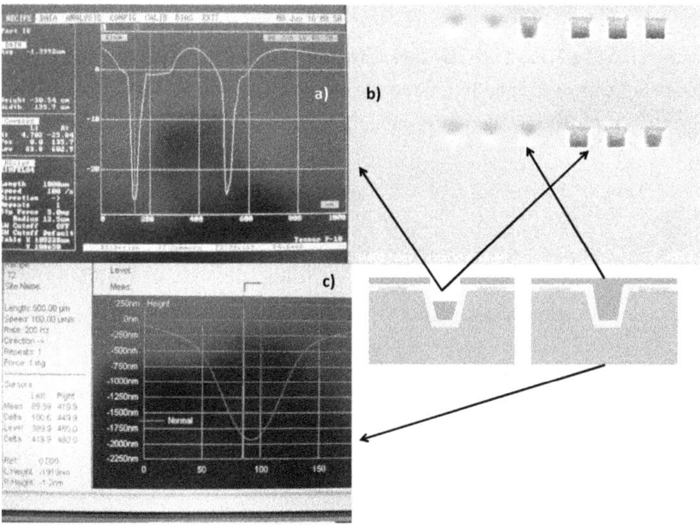

Figure 5.12 Planarization of TSVs. a) bad and c) good example of a screen shot from profiler measurement after BCB application b) SEM micrograph (30° tilted) of planarized TSVs with BCB.

As seen in subframe a) in Fig.5.12, a typical application of BCB with a spin coater is not sufficient. The via hole is not filled entirely. This would lead to a poor wafer bond quality, due to the resulting voids in the TSVs. In the subframe b), the difference between well and poor planarized vias is pointed shown side by side. While groups have reported successful polymer filling with different techniques, only one was expedient [80–82]. In our case the successful proceeding was a two-step BCB application with a low and a high viscosity BCB, respectively. The key to the conformal BCB fill was the utilization of the prewetting technique [81].
Expedient approach after silicon nitride deposition:

- application of BCB 3022 – 35 (low viscosity) with a dedicated thickness on a plain substrate of 2.6 μm with an adhesion promoter (AP3000) and a hard cure at 240 °C for 90 min.
- prewetting the cured BCB with Mesitylene (solvent of BCB and ingredient in BCB before curing).
- application of BCB 3022 – 57 (high viscosity) with a dedicated thickness on a plain substrate of 18 μm without an adhesion promoter and a hard cure at 240 °C for 90 min.

In subframe c) in Fig.5.12, a profiler measurement is shown after the finished planarization. The screen shot shows a remaining trench of a few micrometers. With this shallow trench, the next steps are possible without the danger of occurring voids in the wafer bond process.

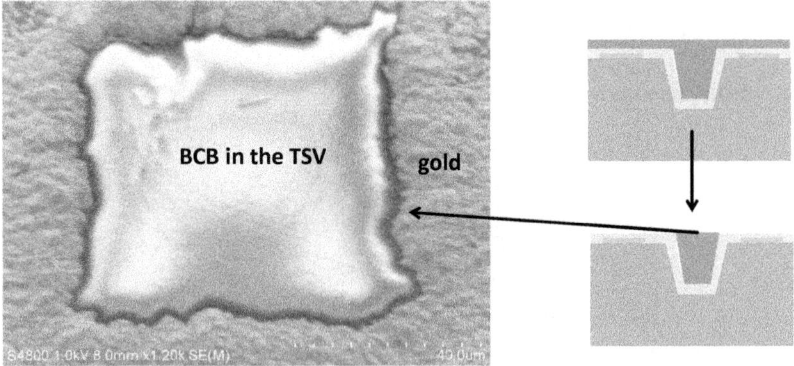

Figure 5.13 SEM micrograph (30° tilted) of etch back the protruding BCB above the gold. Etch recipe: SF_6 - O_2 - 5 - 20 - 100 - 200 - 1.0 at a bias of 340 V with PEEK substrate carrier.

As depicted in Fig.5.13, the back etching of the BCB was successful. The 18 μm BCB were etched with the novel developed etch recipe from Chap. 3.4.4. With the protruding BCB after etching, the next step is the mechanical polishing of the gold until the silicon is reached.

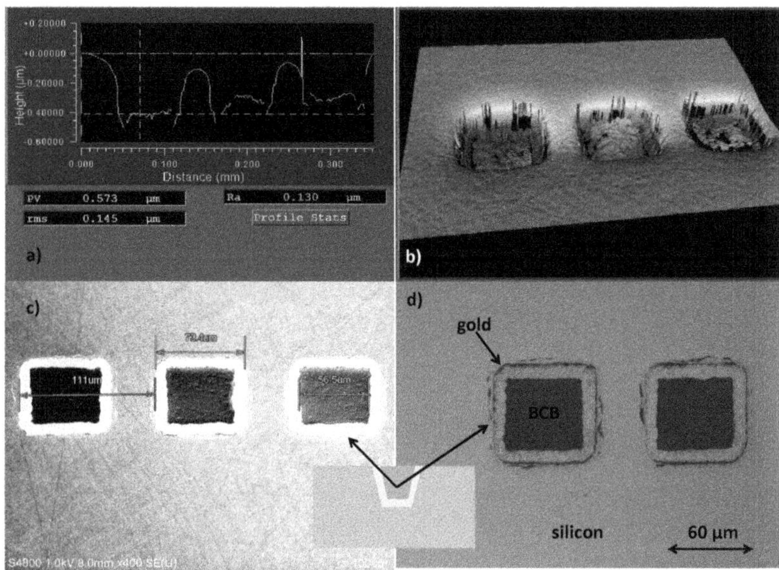

Figure 5.14 TSVs after mechanical polishing. a) optical profiler measurement, b) 3D illustration of the optical profiler measurement c) SEM micrograph (30° tilted) of the polished TSV, d) micrograph of the polished TSVs.

The results after polishing the substrates are shown in Fig.5.14. Subframe a) and b) are depict optical profiler measurements after polishing. The resulting trench indicates a depth of a few hundred nanometers. With such a shallow trench in the TSV, the BCB for wafer bonding is capable to planarize the few hundred nanometers [19]. Subframes c) and d) show SEM and microscope micrographs of the TSVs top surface. The polishing quality shows a flat surface after processing. Comparative measurements with a mechanical profiler showed the same results for the trench depth (all data points within 400 nm). The polishing was performed in two steps, first with a 9 μm silicon carbide slurry for 90 min and the second with 3 μm diamond slurry for 190 min with a force of 1.0 kg for both polishing steps. With the TSV head polished sufficiently, the landing pad on top of the TSV can be applied for electrical measurements (in the actual process for ground connection).

Figure 5.15 TSVs after landing pad lift-off. a) and b) SEM micrographs of the TSV covered with the landing pad, c) micrograph of the landing pad.

Fig.5.15 shows the land pad covering the TSV after lift-off. As metal stack, a mixture of three metals were applied by a dedicated metal evaporator. From TSV to the top, the landing pad consists of titanium (30 nm) for adhesion, gold (200 nm) for increased height and platinum (20 nm) as an effective etch stop for the top connection.

With the applied landing pad, the TSV substrate has to be thinned down to the desired thickness. This step would take place at the end of the process in the actual InP fabrication. Here in the pre-tests, the thinning was necessary to apply the backside metal for electrical measurements and the cross section evaluation of the TSV. The thinning was performed with a 27 μm silicon carbide slurry at 1.5 kg for 150 min.

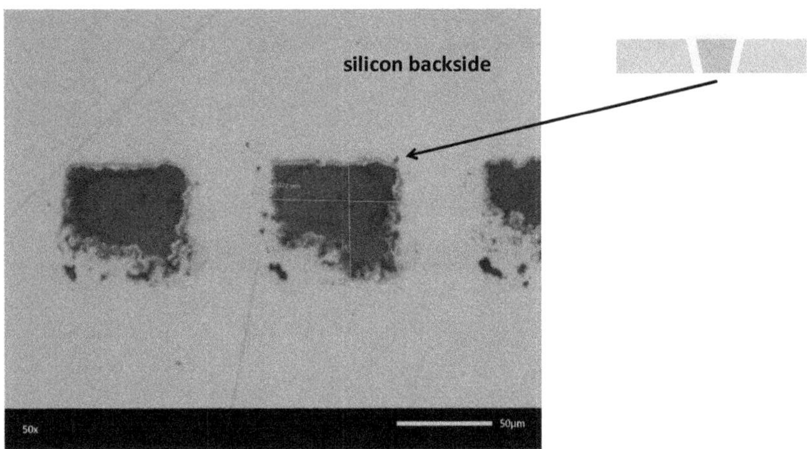

Figure 5.16 Micrograph of the substrate backside after thinning the wafer down to circa 120 μm.

In Fig.5.16, the backside of the TSVs after thinning is shown. The TSVs are clearly visible and indicate a successful via process with gold metal at the edges and the BCB filled inside. The shape is not quadratic anymore, due to the fact that the laser system is out of focus at the via bottom, which degrades the visual quality but not the functionality of the TSV. For suppressing of the parasitic modes, only an electrical connection is necessary, which is not impaired. In the integrated silicon circuits, the TSVs are optimized for low resistance and narrow dimensions (processed with dry etching).

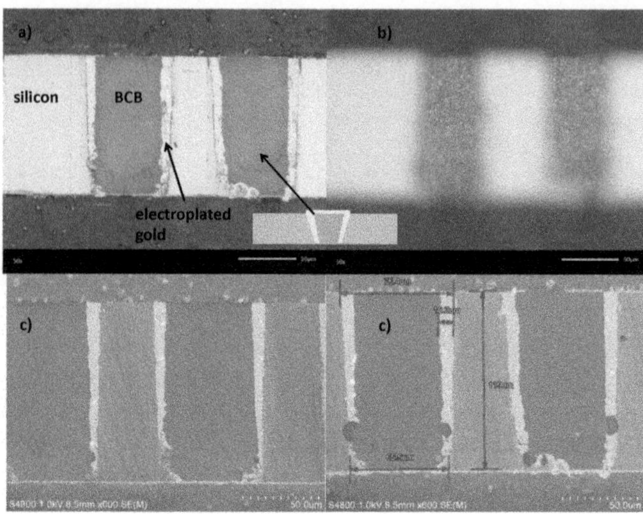

Figure 5.17 Pictures of the TSV cross section. a) and b) microscope micrographs of the TSV, c) and d) SEM micrographs of the TSVs.

In the next step, a cross section of the TSV has been prepared. Fig.5.17 shows a finished TSV after preparing the cross section. First, the micrographs showing a optically functional TSV. With electroplated gold at the via edges and BCB filled inside (viod free), the TSV indicates a functional electrical through substrate connection. At the via bottom, the same characteristic is visible as in Fig.5.16. With the laser system out of focus, the via bottom does not follow the initial via shape, but is electrically functional.

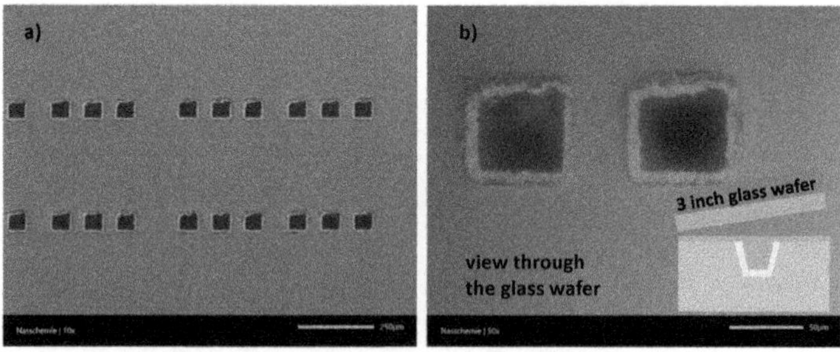

Figure 5.18 TSV substrate after wafer bonding with a glass wafer. a) and b) microscope micrographs of the TSVs with a view through the glass wafer.

Finally, one wafer with TSVs (without the landing pads) was bonded to a glass wafer to evaluate the post-bond quality with the TSV substrate. A wafer with the initial thickness of 700 μm, integrated TSVs was bonded to a glass wafer (3 inch Borofloat 33 substrate) with the actual bond procedure of the InP process. Therefore, 1.5 μm BCB on the silicon and 1.0 μm on the glass side were applied and bonded. The bonding was performed at 240 °C for 90 min with a force of 550 N, similar to the standard process [19]. In Fig.5.18, the post-bond quality is shown in a micrograph. The view is through the glass wafer on the top surface of the TSV wafer. The picture shows an ideal post-bond quality without bubbles that would indicate BCB not filled TSVs. In the next step and last step for the mimic process, the electrical measurements have been done, to evaluate the via connection and the yield.

5.3.3 Electrical results

To determine the electrical functionality of the vias, a thinned wafer with landing pads on the TSVs was completely metalized on the backside of the substrate. The two metals were sputtered with a metal stack of titanium (45 nm) and gold (200 nm), respectively. For measurements, a dedicated process control monitoring layout was used. There, a probe card with 18 pads can be measured with one touchdown of the measurements needles.

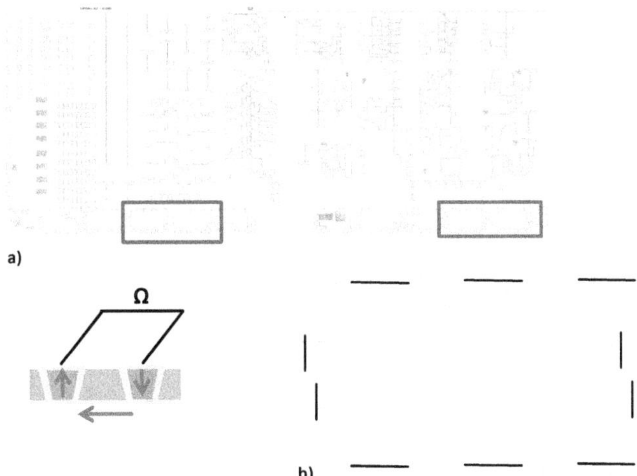

Figure 5.19 Electrical measurements layout. a) layout section of the mimic process with the PCM at the bottom , b) close-up view of a probe card layout with the 10 connections to be measured (red).

In Fig.5.19, the layout of the electrical measurements of the TSVs is shown. In subframe b), the 10 connections of TSVs pairs is shown, which have been measured. For the whole measurement 10 TSV pairs with 4 times per shot were measured. The total amount corresponds to 1020 TSVs pairs and 2040 single TSVs that have been evaluated.

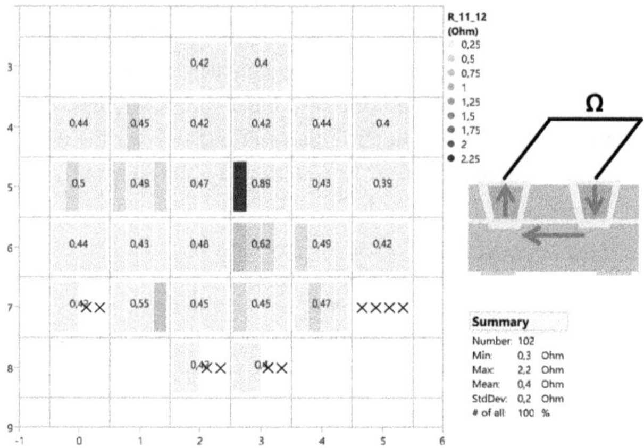

Figure 5.20 Electrical measurements of a first thinned TSV substrate. a) wafer map measurements of a whole 3 inch TSV substrate.

In Fig.5.20, the measurement of the TSV resistance in a wafer map is shown. Here only TSV pairs are shown with a mean value of 0.4 Ω and a standard deviation (S.D.) of 0.2 Ω.

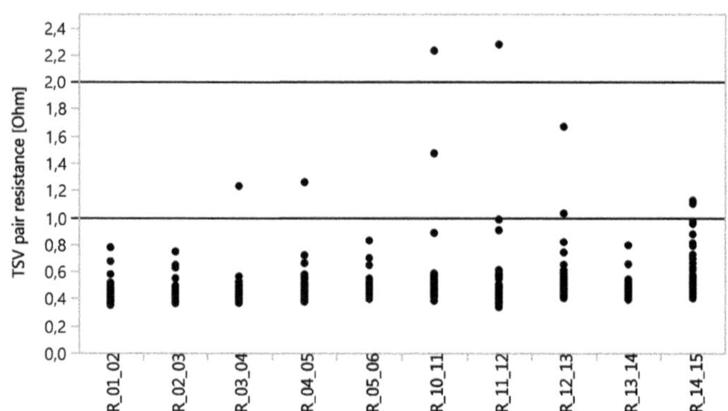

Figure 5.21 Overall measurement data distribution of the TSV pairs.

In total 1020 TSV pairs have been measured. The results are shown in Fig.5.21. The overall mean is 0.633 Ω for the TSV pair. This result in a single via resistance of 0.317 Ω with the neglected series resistance of the backside metal. In total only 6 measurement points were not measurable of 1020 TSV pairs. The measurement discrepancy between Fig.5.21 and Fig.5.20 can be explained with out faded shots with higher resistances (above 2 Ω). With this result, a suitable yield of 99,4% and a low via resistance was proven. Therefore, the laser drilled, gold plated and BCB filled TSV process was ready for implementation in the InP process.

5.4 Implementation into the InP HBT process

Due to the successful mimic process, the TSVs were implemented into the actual InP process. The predefined via layout diameter for the host substrate connection (V1X) is set to 15 μm. Therefore, the design of the TSV layout has been adapted to the provided layout. This is necessary because of the microloading effect described in Chap. 3.2.2. Additionally, a misalignment due to the laser system and wafer bonding was taken into account in the layout design of the TSV. With the a possible laser misalignment in of a few micrometer (as shown in Fig.5.8) and maximum 10 μm in the wafer bonding, the process can cope with a misalignment of more than 10 μm.

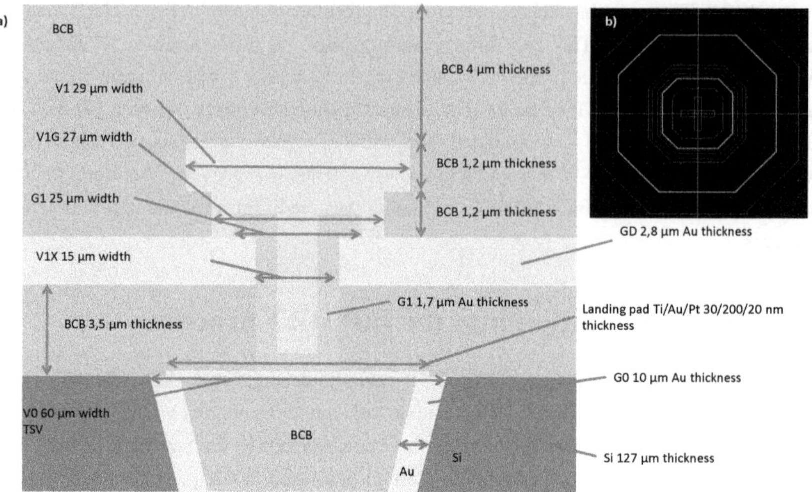

Figure 5.22 Illustration of the TSV layout dedicated for the InP process. a) cross section of the TSV layout, b) top view of the TSV layout.

In Fig.5.22 the layout dimensions for the InP DHBT process are shown. In subframe a), the cross section and in b), the layout top view for the layout design are shown. As seen, the via to the host substrate is designed stepwise. The V1, V1G and V1X are consecutively dry etched, with decreasing via diameter to avoid edge effects during dry etching with simultaneously decreased total etch time.

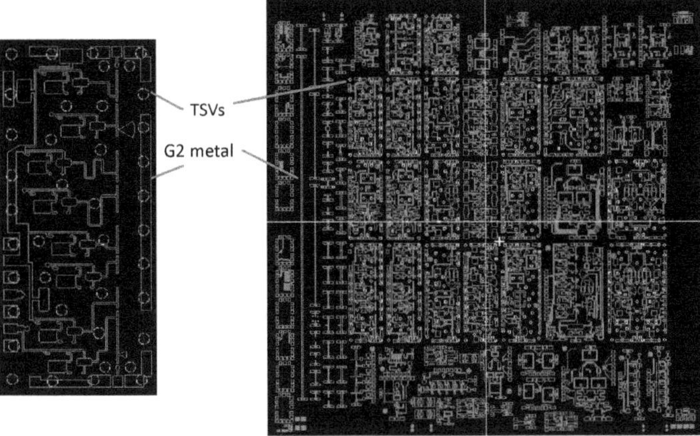

Figure 5.23 Illustration of a InP process layout dedicated for MIMIRAWE project, left) a magnified view if an low noise amplifier (LNA) and right) the complete shot layout is shown.

In Fig.5.23, the layout view of the final design for the MIMIRAWE project is shown. Several circuits such as down- and up converter mixer, low noise amplifier (LNA) and power amplifier (PAs) are implemented. The position of the implemented TSV are shown as dashed white circles in the layout. The solid magenta lines showing the top gold metal (G2) in the InP process.

After the design of the TSV was implemented in the InP design library, a new process run was designed with implemented NiCr resistors (as described in Chap. 4) and the developed TSV process (as described in this Chapter). The process sequence is shown in Fig.5.24. The sequence is besides one detail the same as shown in Fig.5.7. The only difference in the actual process compared with the mimic process, is the position of the alignment marks. In the actual process, the wafer bond alignment has to be performed from the front side of the InP wafer to the backside of the host wafer. Therefore, the alignment marks for the TSV process are on the backside.

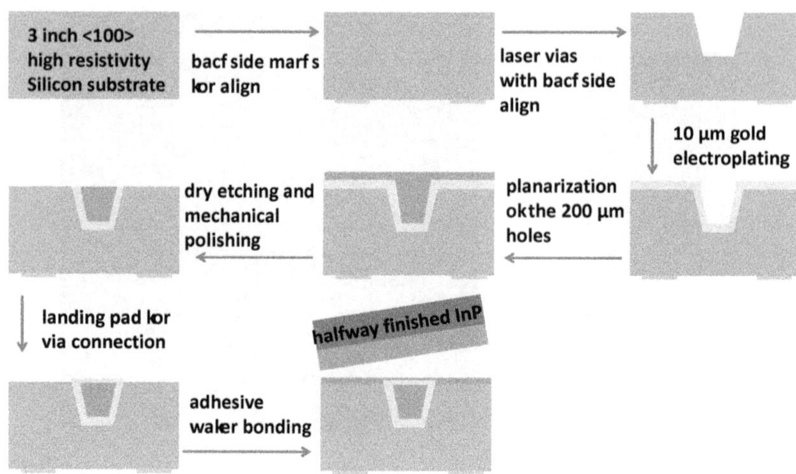

Figure 5.24 Illustration of the TSV process sequence for the InP DHBT process.

In the transferred-substrate technology, a parallel processing of the InP wafer and the host wafer is possible. Therefore, both wafer processes were started in parallel. As first process step, the alignments marks have been applied by the lift-off technique on the backside of the TSV wafers. As layout mask for the marks, the B1 layer from the InP process was chosen, since it has the same mirroring as the InP front side and contains also an alignment cross for wafer bonding.

Due to the lack of an infrared microscope, which could analyze the alignment accuracy of the TSVs to the backside of the silicon wafer, the actual host wafers (TSV wafer) were not checked for alignment after drilling. To determine the laser system alignment accuracy from the backside to the front side, a few glass wafers were processed beforehand, to check the layout mirroring and the alignment accuracy. The results with a misalignment of lower than 5 μm, were similar to the first alignment tests, as already shown in Fig.5.8.

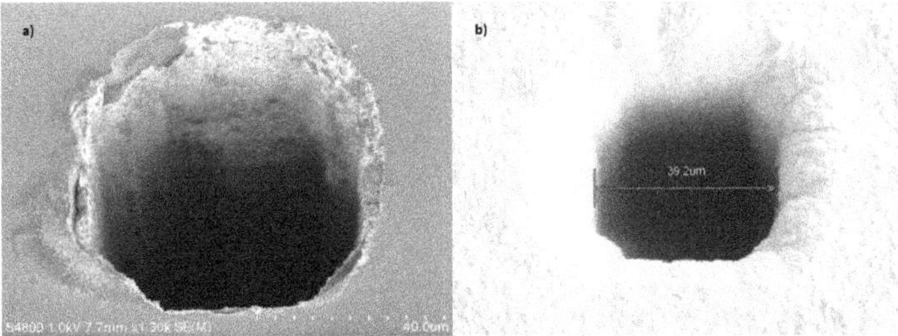

Figure 5.25 SEM micrographs (30° tilted) of laser drilled and electroplated vias. a) after laser drilling and debris cleaning, b) after electroplating the whole wafer with 10 μm gold.

In Fig.5.25, the TSVs are shown after laser drilling and cleaning a) and after electroplating b). As seen, an octagon shape was designed for the TSVs. From the knowledge of BCB etching, an octagonal is a suitable compromise between a square layout and a circle. A square has the drawback of occurring stress at the edges. The circle (with more than 100 sides) is difficult to process due to the large number of polygons which is difficult to handle by the equipment. As seen, the laser drilling and electroplating was successful.

In the next step, the BCB was applied in a two step process as described in Chap. 5.3.2. After the BCB was cured, the BCB was dry etched until the gold surface was reached. Subsequently the gold was polished similar to the mimic process in Chap. 5.3.2.

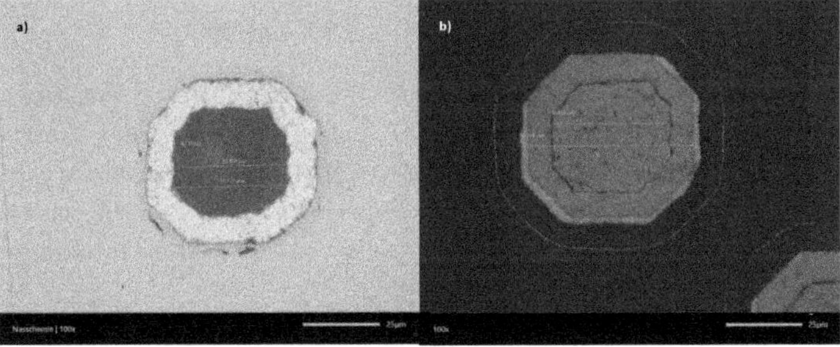

Figure 5.26 Micrographs of polished TSVs and the applied landing pad. a) frontside polishing the TSVs, b) after the landing pad was applied on the top of the TSV.

In Fig.5.26, the TSVs after polishing in subframe a) and landing pad lift-off in subframe b) are shown. In subframe a), a flat and smooth TSV top is visible which is necessary for a high yield wafer bond process. In subframe b), the applied landing pad is shown. After the landing pad was processed, only a 30 nm thick SiNx layer for BCB adhesion was applied. With the host substrate ready for wafer bonding, the dedicated process run with TSV substrates was bonded with InP wafers.

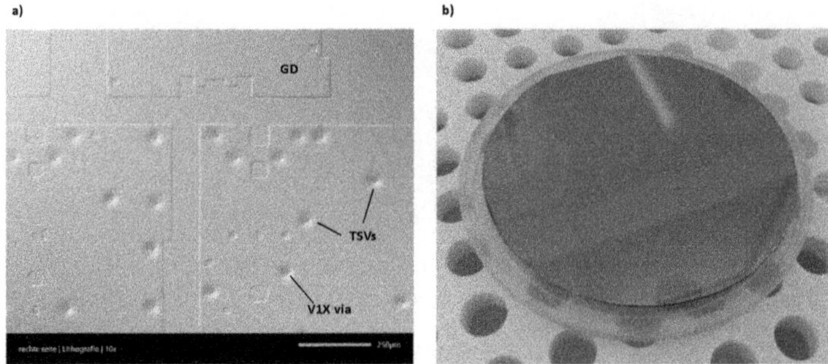

Figure 5.27 Micrographs of bonded wafer after substrate removal. a) chip micrograph of the alignment accuracy, b) photo of the finished 3 inch wafer.

The post bond results are shown in Fig.5.27. In subframe a), the micrograph shows the alignment accuracy of the InP to the host substrate (with TSV). In subframe b), a finished 3 inch wafer after InP substrate removal is shown. In sum, the implementation of the newly developed TSV host substrates was successful. The remaining processing was performed as described in Chap. 2.

5.4.1 InP device measurements with a TSV substrate

After the InP DHBT transferred-substrate process was finished, some preliminary measurements have been performed on wafer. Passives, single HBTs and circuits were measured. Important for the TSV implementation are functional HBTs with no degradation due to TSVs in the host substrates. Therefore the standard HBTs with a width of 0.8 μm and a length of 6 μm were compared. Four wafers in total were compared (two with and two without TSVs).

Table 5.2 On wafer map measurements of finished InP DHBTs.

	without TSVs				with TSVs			
	wafer 1		wafer 2		wafer 3		wafer 4	
	ft	fmax	ft	fmax	ft	fmax	ft	fmax
mean [GHz]	353	355	351	337	341	350	348	332
S.D. [GHz]	2.6	8.1	5.8	12	10.4	5	16.6	8.9

In Tab. 5.2, the comparison of finished InP DHBT wafers are shown. Four wafers were compared. As anticipated, no impact on the device performance is measurable. Wafers with and without TSVs are show the same device performance.

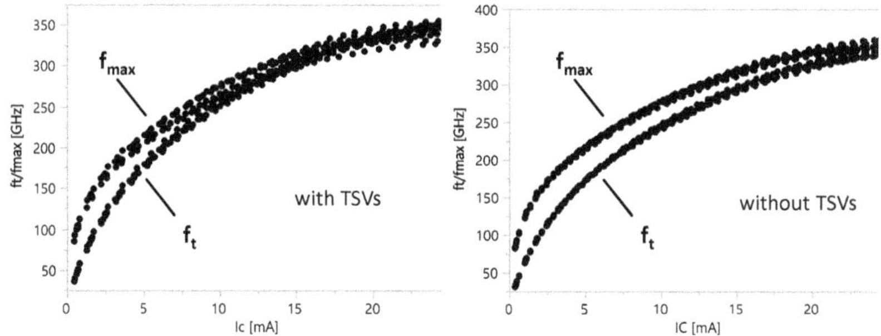

Figure 5.28 f_t and f_{max} vs. Ic measurements of two wafers each with and without TSVs.

In Fig.5.28, the f_t and f_{max} vs. collector current is shown. All wafers are showing the same performance as already pointed out in Tab. 5.2. The final test for functional TSVs is the mounted system into a metal housing, since the parasitic modes only occurring in a mounted and connected system.

With the finished on-wafer measurements, the MMICs have to be diced for mounting. Therefore, saw lanes need to be etched, since the BCB is brittle when its sawed.

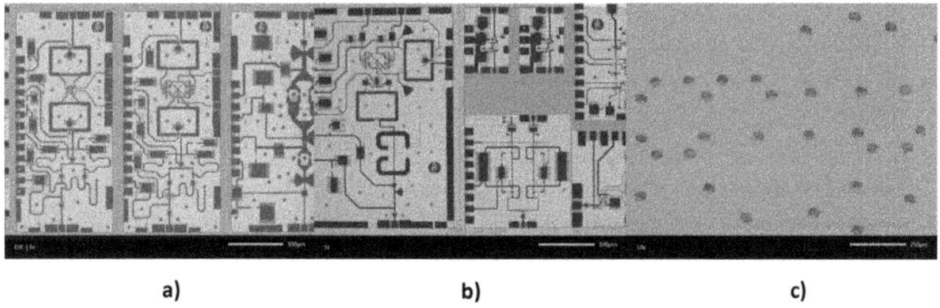

Figure 5.29 Micrographs of the InP MMIC dicing process. a) MMICs covered with photo mask to protect during dry etching, MMICs after dry etching the BCB down to the silicon substrate, c) backside of the MMICs with TSVs sticking out of the silicon substrate.

Fig.5.29 shows the process for MMIC dicing. Subframe a) and b) are showing the saw lanes between the circuits after mask application and after dry etching, respectively. In subframe c), the TSVs are depicted after thinning down the silicon substrate to 127 µm. As dry etch mask, a 17 µm thick AZ 15nxT negative photo resist was used to mask the circuits during the 13 μm BCB dry etching. As dry etch recipe, the newly developed dry etch process for BCB was used, as descried in Chap. 3. Thinning and backside metalization was performed as described in Chap. 5.3.2.

After the saw lane etching and thinning, the MMICs were diced with a commercially available Disco DAD-321 wafer dicer with a diamond saw blade thickness of 60 μm. After the final dicing step, the MMICs have been picked and subsequently shipped for external mounting into the radar module demonstrator.

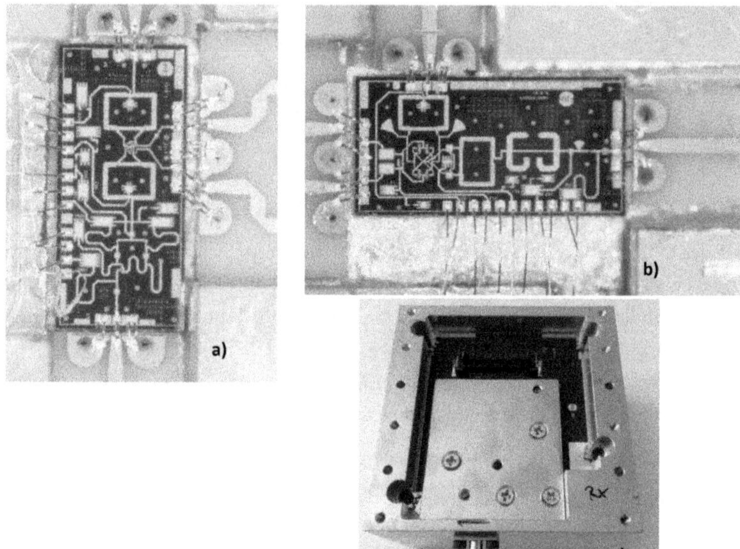

Figure 5.30 Micrographs of the InP MMIC dicing process. a) MMICs covered with photo mask to protect during dry etching, MMICs after dry etching the BCB down to the silicon substrate, c) backside of the MMICs with TSVs sticking out of the silicon substrate.

In Fig.5.30, the mounted MMICs in the receiver modules are depicted. Subframe a) and b) are showing a mixer and down-mixer, respectively. In subframe c), the receiver backside is shown. Transmitter and receiver module was assembled and is now pending for measurement.

5.5 Outlook of the TSVs in the InP process

An advanced method to fabricate TSVs is the utilization of either the BOSCH or Cryo process. The FBH has already ordered a plasma etcher with the capability to fabricate TSVs with both methods. First experiments on a similar plasma etch equipment have been performed.

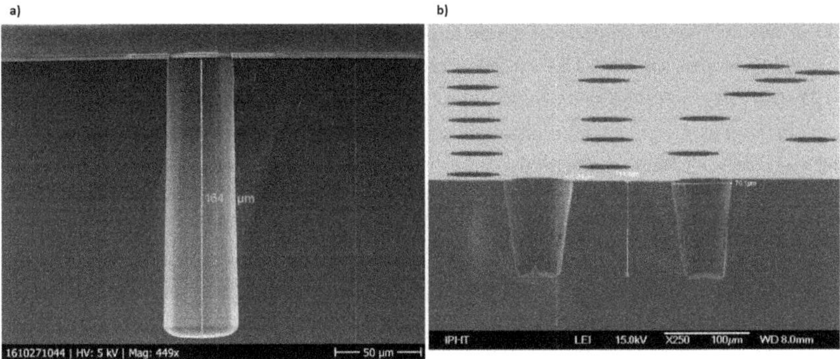

Figure 5.31 Cross section SEM micrographs of dry etch TSVs. a) dry etched with the BOSCH process (performed at Sentech Instruments GmbH), b) dry etched with Cryo process (performed at Leibniz Institute of Photonic Technology's).

As depicted in Fig.5.31, the fabrication of TSVs can be performed either with the BOSCH or Cryp process. Due to the process characteristics, the BOSCH process has straight side walls and the Cryo a variable angle as shown in a) and b), respectively. For the plating base, electroplating and BCB filling, a positive side wall angle like in subframe b), has advantages against the straight BOSCH angle. Therefore the Cryo process will most likely substitute the laser drilling system in the future. With the variable via angle, faster processing, good alignment accuracy and the parallel etching of the TSVs, the Cryo etch process is superior over the laser drilling and will be implemented into the process.

5.6 Chapter 5 summary

In Chap.5, the motivation, the process development, the implementation and the results of the TSVs in the InP transferred-substrate technology were shown. In the motivation, the occurring parasitic parallel plate modes and their impact on the circuit performance were explained. In the process development section, the process development was described step-by-step. The successful laser drilling of silicon showed the potential of such a system for usage in an InP DHBT process. The difficulties in planarizing of deep TSVs with BCB were solved by a two step process. The BCB was applied with a low viscosity layer cured and applied again with a high viscosity. The key to success was the prewetting of the substrate with the BCB solvent (Mesitylene). The electrical measurements of the TSVs have shown functional through substrate connections with circa 0.3 Ω for one via connection. The implementation of TSVs in the actual process was successful. Preliminary high frequency measurements show no degradation of the device performance with TSVs in the host substrate. In conclusion, only with the TSVs implemented in the host substrates, a successful mounting of MMICs into a system is possible. The MMICs with integrated TSVs were diced and implemented in a receiver and transmitter radar module. With this process development, the InP transferred-substrate technology is now commercially usable in a system.

Chapter 6

Summary

To pave the way for higher bandwidth and new applications such as non-invasive blood glucose measurements electronics operating and emitting at frequencies above 300 GHz are needed. A promising candidate for such a task is the InP transferred-substrate process.
In this dissertation the advancement of the InP transferred-substrate process was shown. Therefore, separate topics were discussed, which were split into three chapters (Chap.3, Chap.4, Chap.5). Each topic shows an advancement in the process to either accelerate the process, to enable new features or even facilitate the usage in a system.
In Chap.3, the development of a new dry etch plasma process was described. Since the whole vertical InP HBT is embedded in BCB, the necessity for faster and redundant dry etch process was shown. In this work the newly developed process is capable of fivefold etch rate compared with the predecessor, at maintained bias voltage and anisotropy. Additionally, a formation of an etch by-product was discovered, that originates presumably form the reactor construction (ICP coupler plate). The by-product, that consists of AlF_xO_y was removable with a TMAH-based MF-26A developer. The newly developed ICP process substituted the predecessor RIE process for the fabrication of InP HBTs at the FBH.
Chap.4 deals with the implementation of NiCr TFRs for utilization in the InP HBT process. Highly resistive elements are essential for several circuit topologies. Three different process schemes were evaluated, of which one (bottom contact, structured by lift-off) performed with benchmark contact resistance from NiCr to metal ($8 \times 10^{-10}\ \Omega \cdot cm^2$). Difficulties in processing and electrical performance were identified and disucssed for two technology schemes (bottom contact and top contact, structured by wet etch). Simulated and measured thermal stress, illustrated a poor thermal conduction of NiCr to the heat sink (substrate). Nevertheless, the high frequency measurements of passive and active circuit elements with NiCr embedded, demonstrated the irreplaceable NiCr TFRs in InP MMICs in perfect agreement between simulation and measurement.

In Chap.5, the third and last process advancement in this dissertation is shown. The implementation of TSVs inside the host substrate is an essential element in the transferred-substrate process, to enable an implementation of InP MMICs into complex electronic systems (e.g. radar module). Parasitic substrate modes have motivated with simulations the necessity of an effective method to suppress occurring HF modes in applications. A cheap and reliable method was show to fabricate TSVs and implemented into the InP TS process by the utilization of laser drilling. A process was developed with laser drilled, gold metalized and BCB filled vias. The implementation showed a suitable via resistance with 0.3 Ω for one via through connection. The DC and HF comparison of wafers with and without TSVs, demonstrated no degradation of the device performance with TSVs after fabrication and on-wafer measurement. MMICs with implemented TSVs have been diced and assembled into a radar module, which is now pending for measurement.
In conclusion, the process development in this dissertation shows three key aspects that are now part of the InP transferred-substrate process, to pave the way for future terahertz projects and applications.

List of Figures

List of Tables

Bibliography

[1] P. Siegel et al., "First millimeter-wave animal in vivo measurements of L-Glucose and D-Glucose: Further steps towards a non-invasive glucometer," *International Conference on Infrared, Millimeter, and Terahertz Waves, IRMMW-THz*, vol. 2016-Novem, pp. 1–3, 2016.

[2] P. H. Siegel, "Terahertz technology in biology and medicine," *IEEE Transactions on Microwave Theory and Techniques*, vol. 52, no. 10, pp. 2438–2447, 2004.

[3] J. Schellenberg, B. Kim, and T. Phan, "W-band, broadband 2W GaN MMIC," *IEEE MTT-S International Microwave Symposium Digest*, pp. 59–62, 2013.

[4] M. Urteaga, Z. Griffith, T. Oaks, T. Oaks, and S. Barbara, "InP HBT Technologies for THz Integrated Circuits," vol. 105, no. 6, pp. 1051–1067, 2017.

[5] A. Fox, B. Heinemann, H. Rücker, R. Barth, G. G. Fischer, C. Wipf, S. Marschmeyer, K. Aufinger, J. Böck, S. Boguth, H. Knapp, R. Lachner, W. Liebl, D. Manger, T. F. Meister, A. Pribil, and J. Wursthorn, "Advanced heterojunction bipolar transistor for half-THz SiGe BiCMOS technology," *IEEE Electron Device Letters*, vol. 36, no. 7, pp. 642–644, 2015.

[6] X. Mei, W. Yoshida, M. Lange, J. Lee, J. Zhou, P. H. Liu, K. Leong, A. Zamora, J. Padilla, S. Sarkozy, R. Lai, and W. R. Deal, "First Demonstration of Amplification at 1 THz Using 25-nm InP High Electron Mobility Transistor Process," *IEEE Electron Device Letters*, vol. 36, no. 4, pp. 327–329, 2015.

[7] Y. Shiratori, T. Hoshi, and N. Kashio, "Indium Phosphide-based Heterojunction Bipolar Transistors with Metal Subcollector Fabricated Using Substrate-transfer Technique,"

[8] N. G. Weimann et al., "SciFab -a wafer-level heterointegrated InP DHBT/SiGe BiCMOS foundry process for mm-wave applications," *Physica Status Solidi (A) Applications and Materials Science*, vol. 213, no. 4, pp. 909–916, 2016.

[9] V. Radisic, C. Monier, K. K. Loi, A. Gutierrez-aitken, and S. Member, "W-Band InP HBT Power Amplifiers," vol. 27, no. 9, pp. 824–826, 2017.

[10] K. M. Leong, X. Mei, W. H. Yoshida, A. Zamora, J. G. Padilla, B. S. Gorospe, K. Nguyen, and W. R. Deal, "850 GHz Receiver and Transmitter Front-Ends Using InP HEMT," *IEEE Transactions on Terahertz Science and Technology*, vol. 7, no. 4, pp. 466–475, 2017.

[11] R. Lövblom, R. Flückiger, M. Alexandrova, O. Ostinelli, C. R. Bolognesi, and A. T.-i. Inp, "InP / GaAsSb DHBTs with Simultaneous," vol. 34, no. 8, pp. 984–986, 2013.

[12] J. C. Rode, H.-w. Chiang, P. Choudhary, V. Jain, B. J. Thibeault, W. J. Mitchell, M. J. W. Rodwell, M. Urteaga, D. Loubychev, A. Snyder, Y. Wu, J. M. Fastenau, and A. M. Y. W. K. Liu, "An InGaAs / InP DHBT With Simultaneous f τ / f max 404 / 901 GHz and 4 . 3 V Breakdown Voltage," vol. 3, no. 1, pp. 54–57, 2015.

[13] S. V. Huylenbroeck, A. Sibaja-Hernandez, R. Venegas, S. You, F. Vleugels, D. Radisic, W. Lee, W. Vanherle, K. D. Meyer, and S. Decoutere, "Pedestal Collector Optimization for High Speed SiGe : C HBT," *IEEE Bipolar/BiCMOS Circuits and Technology Meeting*, pp. 66–69, 2011.

[14] N. Weimann et al., "Tight Focus Toward the Future: Tight Material Combination for Millimeter-Wave RF Power Applications: InP HBT SiGe BiCMOS Heterogeneous Wafer-Level Integration," *IEEE Microwave Magazine*, vol. 18, pp. 74–82, mar 2017.

[15] Z. M. Griffith, "Ultra High Speed InGaAs/InP DHBT Devices and Circuits," no. September, 2005.

[16] D. Stoppel, I. Ostermay, M. Hrobak, T. Shivan, M. Hossain, M. Reiner, N. Thiele, K. Nosaeva, M. Brahem, V. Krozer, S. Boppel, N. Halder, and N. Weimann, "NiCr resistors for Terahertz applications in an InP DHBT process," *Microelectronic Engineering*, 2018.

[17] DOW, "CYCLOTENE ™ 3000 Series Advanced Electronics Resins," 2012.

[18] T. Kraemer, "High-Speed InP Heterojunction Bipolar Transistors and Integrated Circuits in Transferred Substrate Technology," 2010.

[19] I. Ostermay, a. Thies, T. Kraemer, W. John, N. Weimann, F.-J. Schmückle, S. Sinha, V. Krozer, W. Heinrich, M. Lisker, B. Tillack, and O. Krüger, "Three-dimensional InP-DHBT on SiGe-BiCMOS integration by means of Benzocyclobutene based wafer bonding for MM-wave circuits," *Microelectronic Engineering*, vol. 125, pp. 38–44, 2014.

[20] N. G. Weimann, S. Monayakul, S. Sinha, F. J. Schmuckle, M. Hrobak, D. Stoppel, W. John, O. Kruger, R. Doerner, B. Janke, V. Krozer, and W. Heinrich, "Manufacturable low-cost flip-chip mounting technology for 300-500-GHz assemblies," *IEEE Transactions on Components, Packaging and Manufacturing Technology*, vol. 7, no. 4, pp. 494–501, 2017.

[21] K. Scherpinski et al., "Integration of NiCr Resistors in a Multilayer CdBCB Wiring System," 1999.

[22] M. Urteaga et al., "THz bandwidth InP HBT technologies and heterogeneous integration with Si CMOS," *Proceedings of the IEEE Bipolar/BiCMOS Circuits and Technology Meeting*, vol. 2016-Novem, pp. 35–41, 2016.

[23] E. Perret, N. Zerounian, S. David, and F. Aniel, "Complex permittivity characterization of benzocyclobutene for terahertz applications," *Microelectronic Engineering*, vol. 85, no. 11, pp. 2276–2281, 2008.

[24] H. Dylla, D. Hoffman, B. Singh, and J. H. Thomas, *Handbook of Vacuum Science and Technology*. No. October, 1997.

[25] W. John, "Apprentices preparation at FBH," 2015.

[26] S. Jensen and O. Hansen, "Characterization of the microloading effect in deep reactive ion etching of silicon," *Micromachining and Microfabrication*, no. Mic, pp. 111–118, 2004.

[27] S. O. Neil, S. Tiku, C. Luo, and P. Zampardi, "Oxygen Plasma Damage Study on InGaP / GaAs HBTs," *CS ManTech*, 2005.

[28] D. Stoppel et al., "Benzocyclobutene dry etch with minimized byproduct redeposition for application in an InP DHBT process," *Microelectronic Engineering*, vol. 161, pp. 63–68, aug 2016.

[29] H. Stieglauer, T. Wiedenmann, H. Bretz, H. Mietz, D. Traulsen, and D. Behammer, "Via etching in BCB for HBT technology," *Semiconductors*, 2008.

[30] Q. Chen, Z. Wang, Z. Tan, and L. Liu, "Characterization of reactive ion etching of benzocyclobutente in SF6 / O2 plasmas," *Microelectronic Engineering*, vol. 87, no. 10, pp. 1945–1950, 2010.

[31] P. Chinoy and J. Tajadod, "Processing and microwave characterization of multilevel interconnects using benzocyclobutene dielectric," *IEEE Transactions on Components, Hybrids, and Manufacturing Technology*, vol. 16, no. 7, pp. 714–719, 1993.

[32] J. Almerico, S. Ross, P. Werbaneth, J. Yang, and P. Garrou, "Plasma etching of thick BCB polymer films for flip chip bonding of hybrid compound semiconductor-silicon devices," *Tegal Crop*, vol. m, 2001.

[33] P. Chinoy, "Reactive ion etching of benzocyclobutene polymer films," *IEEE Transactions on Components, Packaging, and Manufacturing Technology: Part C*, vol. 20, no. 3, pp. 199–206, 1997.

[34] E. Liao, W. Teh, K. Teoh, A. Tay, H. Feng, and R. Kumar, "Etching control of benzocyclobutene in CF4 / O2 and SF6 / O2 plasmas with thick photoresist and titanium masks," *Thin Solid Films*, vol. 504, no. 1-2, pp. 252–256, 2006.

[35] M. Schier, "Reactive Ion Etching of Benzocyclobutene Using a Silicon Nitride Dielectric Etch Mask," *Journal of The Electrochemical Society*, vol. 142, no. 9, p. 3238, 1995.

[36] H. Zhu, J. He, and B. C. Kim, "Processing and characterization of dry-etch benzocyclobutene as substrate and packaging material for neural sensors," *IEEE Transactions on Components and Packaging Technologies*, vol. 30, no. 3, pp. 390–396, 2007.

[37] H. Young-tack, Y. il Kim, M.-c. Lee, P. Sunhee, S. Dongha, C. Park, B. Hong, Y. Roh, S. hae Jung, and I. Song, "Post-etch residue removal in BCB/Cu interconnection structure," *Thin Solid Films*, vol. 435, pp. 238–241, jul 2003.

[38] R. Boucher, "Etching of sub-micron high aspect ratio holes in oxides and polymers," *Microelectronic Engineering*, vol. 73-74, pp. 330–335, 2004.

[39] H. Zhu, J. He, and B. Kim, "Surface-Micromachined Neural Sensors with Integrated Double Side Recordings on Dry-Etch Benzocyclobutene(BCB) Substrate.," *Annual International Conference of the IEEE Engineering in Medicine and Biology Society. IEEE Engineering in Medicine and Biology Society*, vol. 1, pp. 526–9, 2005.

[40] Y. Yang, Z. Song, Y. Du, and Z. Wang, "Reactive ion etching characteristics of partially-cured benzocyclobutene," pp. 862–866, 2014.

[41] G. S. Kim and C. Steinbruechel, "Plasma etching of benzocyclobutene in CF4 O2 and SF6 O2 plasmas," *Journal of Vacuum Science & Technology A: Vacuum, Surfaces, and Films*, vol. 24, no. 3, p. 424, 2006.

[42] M. W. Kirt R. Williams, Kishan Gupta, "Etch Rates for Micromachining Processing Part II," *Journal of Chemical Information and Modeling*, vol. 53, no. 9, pp. 1689–1699, 2013.

[43] Microchemicals, "Aluminium Etching (Al)," 2013.

[44] R. Unger, "Untersuchungen zum BCB-Ätzen mit ICP mit verschiedenen Faradayschirmvarianten an Ätzer4," *FBH Berlin*, 2017.

[45] I. P. Ganachev, M. Moriyama, D. Ogawa, and K. Nakamura, "Faraday Shielding of One-turn Planar ICP Antennas," pp. 8–11, 2016.

[46] S. J. Horst, "Low Cost Fabrication Techniques for Embedded Resistors on Flexible Organics At Millimeter Wave Frequencies Low Cost Fabrication Techniques for Embedded Resistors on Flexible Organics At," no. December, p. 63, 2006.

[47] D. Nachrodt, "Temperaturstabile Dünnfilmwiderstände aus Ti/TiN und Ti/NiCr mit niedrigem Temperaturkoeffizienten und ihre Integration in einen Standard-CMOS-Prozess," 2008.

[48] H. Wheeler, "Formulas for the Skin Effect," *Proceedings of the IRE*, vol. 30, no. 9, pp. 412–424, 1942.

[49] Fridman, "Contact resistance and TLM measurements," pp. 1–11, 2014.

[50] Microwaves101, "Van der Pauw Measurements," pp. 1–6, 2018.

[51] R. Kopf et al., "Thin-film resistor fabrication for InP technology applications," *Journal of Vacuum Science & Technology B*, vol. 20, no. 3, pp. 871–875, 2002.

[52] J. Rolke et al., "Nichrome Thin Film Technology and its Application," *Electrocomponent Science and Technology*, vol. 9, no. C, pp. 51–57, 1981.

[53] R. Driad et al., "Investigation of NiCr Thin Film Resistors for InP-Based Monolithic Microwave Integrated Circuits (MMICs)," *Journal of The Electrochemical Society*, vol. 158, no. 5, pp. H561–H564, 2011.

[54] C. H. Lane, "Nichrome resistor properties and reliability," pp. Report RADC–TR–73–181, 6/73, 1973.

[55] Z. Huang, N. Geyer, P. Werner, J. De Boor, and U. Gösele, "Metal-assisted chemical etching of silicon: A review," *Advanced Materials*, vol. 23, no. 2, pp. 285–308, 2011.

[56] S. Yae, Y. Morii, N. Fukumuro, and H. Matsuda, "Catalytic activity of noble metals for metal-assisted chemical etching of silicon," *Nanoscale Research Letters*, vol. 7, pp. 1–14, 2012.

[57] E. Schmucker et al., "Oxidation of Ni-Cr alloy at intermediate oxygen pressures. I. Diffusion mechanisms through the oxide layer," *Corrosion Science*, vol. 111, pp. 474–485, 2016.

[58] E. Bloch et al., "NiCr thin film resistor integration with InP technology," *Semiconductor Science and Technology*, vol. 26, no. 10, p. 105004, 2011.

[59] W. J. Roesch, "Physical degradation characterization of thin film resistors," *2007 ROCS Workshop (formerly the GaAs REL Workshop)- Reliability of Compound Semiconductors, Proceedings*, no. 503, pp. 103–114, 2007.

[60] Massachusetts Institute of Technology, "Thermal coductivity of Gold."

[61] I. HSM Wire International, "Nichrome 80/20," p. 2013, 2013.

[62] SWI, "AlN Thermal Coductivity."

[63] J. Yang et al., "Low Energy Sputter Deposition and Properties of NiCr Thin Film Resistors for GaAs Integrated Circuits," *Structure*, vol. 2, p. 20.

[64] K. Bansal et al., "Design and Analysis of Thin Film Nichrome Resistor for GaN MMICs," pp. 2–5, 2016.

[65] G. Engen et al., "Thru-Reflect-Line: An Improved Technique for Calibrating the Dual Six-Port Automatic Network Analyzer," *IEEE Transactions on Microwave Theory and Techniques*, vol. 27, pp. 987–993, dec 1979.

[66] T. Shivan et al., "A Highly Efficient Ultra-Wideband Travelling- Wave Amplifier in InP DHBT Technology," *MWCL-07092017*, pp. 2–4.

[67] M. Hrobak, *Critical mm-Wave Components for Synthetic Automatic Test Systems*. Springer, 2015.

[68] J. Moll et al., "Panel Design of a MIMO Imaging Radar at W-Band for Space Applications," in *EuRAD08-3*, 2017.

[69] S. Maas, "Ballasting HBTs for wireless power amplifier operation," *2006 International Workshop on Integrated Nonlinear Microwave and Millimeter-Wave Circuits, INMMIC 2006 - Proceedings*, vol. 630, pp. 2–5, 2007.

[70] T. Shivan et al., "A 95 GHz Bandwidth 12 dBm Output Power Distributed Amplifier in InP-DHBT Technology for Optoelectronic Applications," *accepted for publication at GEMIC2018. Freiburg, Germany*.

[71] M. Hrobak, "Internal Insitute Presentation about progress in MIMIRAWE," no. April, 2016.

[72] Teledyne, "Data sheet for Solid state Power Amplifier in 250nm InP HBT," *Data sheet*.

[73] E. Jung, A. Ostmann, P. Ramm, J. Wolf, M. Toepper, and M. Wiemer, "Through silicon vias as enablers for 3D systems," *DTIP of MEMS and MOEMS - Symposium on Design, Test, Integration and Packaging of MEMS/MOEMS*, no. April, pp. 119–122, 2008.

[74] W. Commons, "Anisotropic Wet Etching," 2010.

[75] L. Li, J. Wu, and C. P. Wong, "Wafer-level wet etching of high-aspect-ratio through silicon vias (TSVs) with high uniformity and low cost for silicon interposers with high-density interconnect of 3D packaging," *Proceedings - Electronic Components and Technology Conference*, vol. 2015-July, pp. 1417–1422, 2015.

[76] FBH, "FBH Homepage Process department," pp. 1–2, 2018.

[77] D. Stoppel, M. Hrobak, A. Külberg, S. Schönfeld, D. Rentner, O. Krüger, K. Nosaeva, M. Brahem, S. Boppel, and N. Weimann, "Through silicon via (TSV) process for suppression of substrate modes in a transferred-substrate InP DHBT MMIC technology (abstract for CSW 2018)," 2018.

[78] K. Zdansky, L. Pekarek, and P. Hlidek, "Electronic and optical properties of copper doped InP single crystals," *International Conference on Indium Phosphide and Related Materials, 2005.*, pp. 171–174, 2005.

[79] DOW, "Adhesions of CYCLOTENE Advanced Electronics Resin," pp. 1–13, 2012.

[80] R. Trichur, M. Fowler, J. McCutcheon, and M. Daily, "Filling and Planarizing Deep Trenches with Polymeric Material for Through-Silicon Via Technology," *International Symposium on Microelectronics*, pp. 192–196, 2010.

[81] F. F. Duval, C. Okoro, Y. Civale, P. Soussan, and E. Beyne, "Polymer filling of silicon trenches for 3-D through silicon vias applications," *IEEE Transactions on Components, Packaging and Manufacturing Technology*, vol. 1, no. 6, pp. 825–832, 2011.

[82] Y. Ding, Y. Yan, Q. Chen, S. Wang, X. Chen, and Y. Chen, "Investigation on mechanism of polymer filling in high-aspect-ratio trenches for through-silicon-via (TSV) application," *Science China Technological Sciences*, vol. 57, no. 8, pp. 1616–1625, 2014.

Appendix A

Publications

- "Benzocyclobutene dry etch with minimized byproduct redeposition for application in an InP DHBT process", **D. Stoppel**, W. John, U. Zeimer, K. Kunkel, M.I. Schukfeh, O. Krüger, N. Weimann, Microelectron. Eng., vol. 161, pp. 63 – 68 (2016).
- "Through silicon via (TSV) process for suppression of substrate modes in a transferred-substrate InP DHBT MMIC technology", **D. Stoppel**, M. Hrobak, A. Külberg, S.Schönfeld, D. Rentner, O.Krüger, K. Nosaeva, M. Brahem, S. Boppel and N. Weimann, Oral paper CSW 2018.
- "NiCr resistors for Terahertz applications in an InP DHBT process", **D.Stoppel**, I. Ostermay, M. Hrobak, T. Shivan, M. Hossain, M. Reiner, N. Thiele, K. Nosaeva, M. Brahem, V. Krozer, S. Boppel, N.G. Weimann, Microelectron. Eng., vol. 208, pp. 1 – 6 (2019).
- "Transferred-Substrate InP/GaAsSb Heterojunction Bipolar Transistor Technology With fmax 0.53 THz", N.G. Weimann, T.K. Johansen, **D. Stoppel**, M. Matalla, M. Brahem, K. Nosaeva, S. Boppel, N. Volkmer, I. Ostermay, V. Krozer, O. Ostinelli, and C.R. Bolognesi, IEEE Trans. Electron Devices, vol. 65, no. 9, pp. 3704 – 3710 (2018).
- "A Hetero-Integrated W-Band Transmitter Module in InP-on-BiCMOS Technology", M. Hossain, M.H. Eissa, M. Hrobak, **D. Stoppel**, N. Weimann, A. Malignaggi, A. Mai, D. Kissinger, W. Heinrich, V. Krozer, 13th European Microwave Integrated Circuits Conference (EuMIC 2018), Madrid, Spain, Sep. 24 – 25, pp. 97 – 100 (2018).

- "An Active High Conversion Gain W-Band Up-Converting Mixer for Space Applications", M. Hossain, M. Hrobak, **D. Stoppel**, W. Heinrich, V. Krozer, IEEE MTT-S Int. Microw. Symp. Dig., Philadelphia, USA, Jun. 10 – 15, pp. 682 – 685 (2018).
- "An Ultra-broadband Low-Noise Distributed Amplifier in InP DHBT Technology", T. Shivan, M. Hossain, **D. Stoppel**, N. Weimann, S. Schulz, R. Doerner, V. Krozer, W. Heinrich, 48th European Microwave Conference (EuMC 2018), Madrid, Spain, Sep. 25 – 27, pp. 1209 – 1212 (2018).
- "A 95 GHz Bandwidth 12 dBm Output Power Distributed Amplifier in InP-DHBT Technology for Optoelectronic Applications", T. Shivan, N. Weimann, M. Hossain, T. Johansen, **D. Stoppel**, S. Schulz, O. Ostinelli, R. Doerner, C.R. Bolognesi, V. Krozer, W. Heinrich, 11th German Microwave Conference (GeMiC 2018), Freiburg, Germany, Mar. 12 – 14, ISBN 978 – 3 – 9812668 – 8 – 7, pp. 17 – 20 (2018).
- "A Highly Efficient Ultra-Wideband Travelling-Wave Amplifier in InP DHBT Technology", T. Shivan, N. Weimann, M. Hossain, **D. Stoppel**, S. Boppel, R. Doerner, V. Krozer and W. Heinrich, IEEE Microwave Wireless Compon. Lett., vol. 28, no. 11, pp. 1029 – 1031 (2018).
- "220 – 325 GHz high-isolation SPDT switch in InP DHBT technology", T. Shivan, M. Hossain, **D. Stoppel**, N. Weimann, S. Boppel, R. Doerner, W. Heinrich and V. Krozer, Electron. Lett., vol. 54, no. 21, pp. 1222 – 1224 (2018).
- "An efficient W-band InP DHBT digital power amplifier", A. Wentzel, M. Hossain, **D. Stoppel**, N Weimann, V. Krozer and W. Heinrich, Int. J. Microwave Wireless Technolog., vol. 9, no. 6, pp. 1241 – 1249 (2017).
- "Manufacturable Low-Cost Flip-Chip Mounting Technology for 300 – 500–GHz Assemblies", N.G. Weimann, S. Monayakul, S. Sinha, F.-J. Schmückle, M. Hrobak, **D. Stoppel**, W. John, O. Krüger, R. Doerner, B. Janke, V. Krozer, and W. Heinrich, IEEE Trans., vol. 7, no. 4, pp. 494 – 501 (2017).
- "SciFab - a wafer-level heterointegrated InP DHBT/SiGe BiCMOS foundry process for mm-wave applications", N.G. Weimann, **D. Stoppel**, M.I. Schukfeh, M. Hossain, T. Al-Sawaf, B. Janke, R. Doerner, S. Sinha, F.-J. Schmückle, O. Krüger, V. Krozer, W. Heinrich, M. Lisker, A. Krüger, A. Datsuk, C. Meliani, and B. Tillack, phys. stat. sol. (a), vol. 213, no. 4, pp.909 – 916 (2016).

- "Multifinger Indium Phosphide Double-Heterostructure Transistor Circuit Technology With Integrated Diamond Heat Sink Layer", K. Nosaeva, T. Al-Sawaf, W. John, **D. Stoppel**, M. Rudolph, F.-J. Schmückle, B. Janke, O. Krüger, V. Krozer, W. Heinrich, and N.G. Weimann, IEEE Trans. Electron Devices, vol. 63, no. 5, pp. 1846 – 1852 (2016).

- "An Efficient W-Band InP DHBT Digital Power Amplifier", A. Wentzel, M. Hossain, **D. Stoppel**, N. Weimann, V. Krozer, and W. Heinrich, Proc. 11th (EuMIC 2016), London, UK, Oct. 3 – 4, pp. 21 – 24 (2016).

- "Process Robustness and Reproducibility of sub-mm Wave Flip-Chip Interconnect Assembly", S. Monayakul, S. Sinha, F.J. Schmückle, M. Hrobak, **D. Stoppel**, O. Krüger, B. Janke, N.G. Weimann, (EPEPS 2015), San Jose, CA, Oct 25 – 28, pp. 141 – 144 (2015).

- Patent: "Verfahren zur Herstellung von integrierten elektronischen Schaltungen" (English.:"Process for fabrication of integrated electronic circuits"), **D. Stoppel**, N. Weimann, M. Hrobak, registered at 21.03.2017, DRN:2017032115230100*DE*

Appendix B

Acronyms

Al	aluminum
Al2O3	aluminum oxide
AlN	aluminum nitride
ARDE	aspect ratio dependent effect
Au	gold
B1	base metal
BCB	benzocyclobutene
BCL	bottom connection with lift-off
BCW	bottom connection with wet etch
BiCMOS	bipolar junction transistor CMOS
C	carbon
C1	capacitor
CD	critical dimension
CH3	methyl group
CMOS	complementary metal-oxide-semiconductor
CMP	chemical mechanical polishing
CNC	computerized numerical control
DC	direct current
Di	deionized water
DRIE	deep reactive ion etching
E1	emitter metal
EDX	energy-dispersive X-ray
EM	electromagnetic
F	fluorine

FBH	Ferdinand-Braun-Institute
G1	first interconnection metal
G1C	capacitor bottom metal
G2	second interconnection metal
GaN	gallium nitride
GD	ground metal
H	hydrogen
H2O2	hydrogen peroxide
HBT	heterobipolar transistor
HCl	hydrochloric acid
He	helium
HEMT	high-electron-mobility transistor
HF	high frequency
ICP	inductive coupled plasma
InGaAs	indium gallium arsenide
InP	indium phosphide
K1	collector metal
KOH	potassium hydroxide
LNA	low noise amplifier
MaCE	metal-assisted chemical etching
MIM	metal-insulator-metal
MMIC	monolithic microwave integrated circuits
mTRL	multiline thru-reflect-line
N2	nitrogen
NH3	ammonia
NiCr	nickel-chrome
NMP	N-Methyl-2-pyrrolidone
O	oxygen
PA	power amplifier
PAE	power added efficiency
PCM	process control monitoring
PECVD	plasma-enhanced chemical vapor deposition
PEEK	polyether ether ketone
R1	first planarization etch
RF	radio frequency
RIE	reactive ion etching

Rx	receiver
S2	collector encapsulation
SB	dicing streets
SD	standard deviation
SEM	scanning electron microscope
SF6	sulfur hexafluoride
Si	silicon
SiF4	tetrafluorosilane
SiGe	silicon germanium
SiNx	silicon nitride
TaN	tantalum nitride
TCR	temperature coefficient
TCW	top connection with wet etch
TFR	thin film resistor
Ti	titanium
TIA	transimpedance amplifier
TiW	titanium tungsten
TLM	transmission line method
TMAH	tetramethylammonium hydroxide
TS	transferred substrate
TSV	through silicon vias
TWA	traveling wave amplifier
Tx	transmitter
UV	ultra violet
V1	first via
V1G	ground via
V1X	substrate via
V2	second via
VCO	voltage-controlled oscillator
VDP	van der Pauw method
W1	resistor metal
W2	handover metal for resistor
TMIC	terahertz microwaveintegrated circuits

Innovationen mit Mikrowellen und Licht
Forschungsberichte aus dem Ferdinand-Braun-Institut, Leibniz-Institut für Höchstfrequenztechnik

Herausgeber: Prof. Dr. G. Tränkle, Prof. Dr.-Ing. W. Heinrich

Band 1: **Thorsten Tischler**
Die Perfectly-Matched-Layer-Randbedingung in der Finite-Differenzen-Methode im Frequenzbereich: Implementierung und Einsatzbereiche
ISBN: 3-86537-113-2, 19,00 EUR, 144 Seiten

Band 2: **Friedrich Lenk**
Monolithische GaAs FET- und HBT-Oszillatoren mit verbesserter Transistormodellierung
ISBN: 3-86537-107-8, 19,00 EUR, 140 Seiten

Band 3: **R. Doerner, M. Rudolph (eds.)**
Selected Topics on Microwave Measurements, Noise in Devices and Circuits, and Transistor Modeling
ISBN: 3-86537-328-3, 19,00 EUR, 130 Seiten

Band 4: **Matthias Schott**
Methoden zur Phasenrauschverbesserung von monolithischen Millimeterwellen-Oszillatoren
ISBN: 978-3-86727-774-0, 19,00 EUR, 134 Seiten

Band 5: **Katrin Paschke**
Hochleistungsdiodenlaser hoher spektraler Strahldichte mit geneigtem Bragg-Gitter als Modenfilter (α-DFB-Laser)
ISBN: 978-3-86727-775-7, 19,00 EUR, 128 Seiten

Band 6: **Andre Maaßdorf**
Entwicklung von GaAs-basierten Heterostruktur-Bipolartransistoren (HBTs) für Mikrowellenleistungszellen
ISBN: 978-3-86727-743-3, 23,00 EUR, 154 Seiten

Band 7: **Prodyut Kumar Talukder**
Finite-Difference-Frequency-Domain Simulation of Electrically Large Microwave Structures using PML and Internal Ports
ISBN: 978-3-86955-067-1, 19,00 EUR, 138 Seiten

Band 8: **Ibrahim Khalil**
Intermodulation Distortion in GaN HEMT
ISBN: 978-3-86955-188-3, 23,00 EUR, 158 Seiten

Band 9: **Martin Maiwald**
Halbleiterlaser basierte Mikrosystemlichtquellen für die Raman-Spektroskopie
ISBN: 978-3-86955-184-5, 19,00 EUR, 134 Seiten

Band 10: **Jens Flucke**
Mikrowellen-Schaltverstärker in GaN- und GaAs-Technologie Designgrundlagen und Komponenten
ISBN: 978-3-86955-304-7, 21,00 EUR, 122 Seiten

Cuvillier Verlag
Internationaler wissenschaftlicher Fachverlag

Innovationen mit Mikrowellen und Licht
Forschungsberichte aus dem Ferdinand-Braun-Institut, Leibniz-Institut für Höchstfrequenztechnik

Herausgeber: Prof. Dr. G. Tränkle, Prof. Dr.-Ing. W. Heinrich

Band 11: **Harald Klockenhoff**
Optimiertes Design von Mikrowellen-Leistungstransistoren
und Verstärkern im X-Band
ISBN: 978-3-86955-391-7, 26,75 EUR, 130 Seiten

Band 12: **Reza Pazirandeh**
Monolithische GaAs FET- und HBT-Oszillatoren
mit verbesserter Transistormodellierung
ISBN: 978-3-86955-107-8, 19,00 EUR, 140 Seiten

Band 13: **Tomas Krämer**
High-Speed InP Heterojunction Bipolar Transistors
and Integrated Circuits in Transferred Substrate Technology
ISBN: 978-3-86955-393-1, 21,70 EUR, 140 Seiten

Band 14: **Phuong Thanh Nguyen**
Investigation of spectral characteristics of solitary
diode lasers with integrated grating resonator
ISBN: 978-3-86955-651-2, 24,00 EUR, 156 Seiten

Band 15: **Sina Riecke**
Flexible Generation of Picosecond Laser Pulses in the Infrared and
Green Spectral Range by Gain-Switching of Semiconductor Lasers
ISBN: 978-3-86955-652-9, 22,60 EUR, 136 Seiten

Band 16: **Christian Hennig**
Hydrid-Gasphasenepitaxie von versetzungsarmen und freistehenden
GaN-Schichten
ISBN: 978-3-86955-822-6, 27,00 EUR, 162 Seiten

Band 17: **Tim Wernicke**
Wachstum von nicht- und semipolaren InAlGaN-Heterostrukturen
für hocheffiziente Licht-Emitter
ISBN: 978-3-86955-881-3, 23,40 EUR, 138 Seiten

Band 18: **Andreas Wentzel**
Klasse-S Mikrowellen-Leistungsverstärker mit GaN-Transistoren
ISBN: 978-3-86955-897-4, 29,65 EUR, 172 Seiten

Band 19: **Veit Hoffmann**
MOVPE growth and characterization of (In,Ga)N quantum structures
for laser diodes emitting at 440 nm
ISBN: 978-3-86955-989-6, 18,00 EUR, 118 Seiten

Band 20: **Ahmad Ibrahim Bawamia**
Improvement of the beam quality of high-power broad area
semiconductor diode lasers by means of an external resonator
ISBN: 978-3-95404-065-0, 21,00 EUR, 126 Seiten

Cuvillier Verlag
Internationaler wissenschaftlicher Fachverlag

Innovationen mit Mikrowellen und Licht
Forschungsberichte aus dem Ferdinand-Braun-Institut, Leibniz-Institut für Höchstfrequenztechnik

Herausgeber: Prof. Dr. G. Tränkle, Prof. Dr.-Ing. W. Heinrich

Band 21: **Agnietzka Pietrzak**
Realization of High Power Diode Lasers with Extremely Narrow Vertical Divergence
ISBN: 978-3-95404-066-7, 27,40 EUR, 144 Seiten

Band 22: **Eldad Bahat-Treidel**
GaN-based HEMTs for High Voltage Operation
Design, Technology and Characterization
ISBN: 978-3-95404-094-0, 41,10 EUR, 220 Seiten

Band 23: **Ponky Ivo**
AlGaN/GaN HEMTs Reliability:
Degradation Modes and Anslysis
ISBN: 978-3-95404-259-3, 23,55 EUR, 132 Seiten

Band 24: **Stefan Spießberger**
Compact Semiconductor-Based Laser Sources
with Narrow Linewidth and High Output Power
ISBN: 978-3-95404-261-6, 24,15 EUR, 140 Seiten

Band 25: **Silvio Kühn**
Mikrowellenoszillatoren für die Erzeugung von atmosphärischen Mikroplasmen
ISBN: 978-3-95404-378-1, 21,85 EUR, 112 Seiten

Band 26: **Sven Schwertfeger**
Experimentelle Untersuchung der Modensynchronisation in Multisegment-Laserdioden zur Erzeugung kurzer optischer Pulse bei einer Wellenlänge von 920 nm
ISBN: 978-3-95404-471-9, 29,45 EUR, 150 Seiten

Band 27: **Christoph Matthias Schultz**
Analysis and mitigation of the factors limiting the effiency of high power distributed feedback diode lasers
ISBN: 978-3-95404-521-1, 68,40 EUR, 388 Seiten

Band 28: **Luca Redaelli**
Design and fabrication of GaN-based laser diodes for single-mode and narrow-linewidth applications
ISBN: 978-3-95404-586-0, 29,70 EUR, 176 Seiten

Band 29: **Martin Spreemann**
Resonatorkonzepte für Hochleistungs-Diodenlaser
mit ausgedehnten lateralen Dimensionen
ISBN: 978-3-95404-628-7, 25,15 EUR, 128 Seiten

Cuvillier Verlag
Internationaler wissenschaftlicher Fachverlag

Innovationen mit Mikrowellen und Licht
Forschungsberichte aus dem Ferdinand-Braun-Institut, Leibniz-Institut für Höchstfrequenztechnik

Herausgeber: Prof. Dr. G. Tränkle, Prof. Dr.-Ing. W. Heinrich

Band 30: **Christian Fiebig**
Diodenlaser mit Trapezstruktur und hoher Brillanz für die Realisierung einer Frequenzkonversion auf einer mikro-optischen Bank
ISBN: 978-3-95404-690-4, 26,30 EUR, 140 Seiten

Band 31: **Viola Küller**
Versetzungsreduzierte AlN- und AlGaN-Schichten als Basis für UV LEDs
ISBN: 978-3-95404-741-3, 34,40 EUR, 164 Seiten

Band 32: **Daniel Jedrzejczyk**
Efficient frequency doubling of near-infrared diode lasers using quasi phase-matched waveguides
ISBN: 978-3-95404-958-5, 27,90 EUR, 134 Seiten

Band 33: **Sylvia Hagedorn**
Hybrid-Gasphasenepitaxie zur Herstellung von Aluminiumgalliumnitrid
ISBN: 978-3-95404-985-1, 38,00 EUR, 176 Seiten

Band 34: **Alexander Kravets**
Advanced Silicon MMICs for mm-Wave Automotive Radar Front-Ends
ISBN: 978-3-95404-986-8, 31,90 EUR, 156 Seiten

Band 35: **David Feise**
Longitudinale Modenfilter für Kantanemitter im roten Spektralbereich
ISBN: 978-3-7369-9116-3, 39,20 EUR, 168 Seiten

Band 36: **Ksenia Nosaeva**
Indium phosphide HBT in thermally optimized periphery for applications up to 300GHZ
ISBN: 978-3-7369-287-0, 42,00 EUR, 154 Seiten

Band 37: **Muhammad Maruf Hossain**
Signal Generation for Millimeter Wave and THZ Applications in InP-DHBT and InP-on-BiCMOS Technologies
ISBN: 978-3-7369-9335-8, 35,60 EUR, 136 Seiten

Band 38: **Sirinpa Monayakul**
Development of Sub-mm Wave Flip-Chip Interconnect
ISBN: 978-3-7369-9410-2, 44,00 EUR, 146 Seiten

Band 39: **Moritz Brendel**
Charakterisierung und Optimierung von (Al, Ga) N-basierten UV-Photodetektoren
ISBN: 978-3-7369-9465-2, 49,90 EUR, 196 Seiten

Band 40: **Erdenetsetseg Luvsandamdin**
Development of micro-integrated diode lasers for precision quantum optics experiments in space
ISBN: 978-3-7369-9479-9, 39,00 EUR, 126 Seiten

Cuvillier Verlag
Internationaler wissenschaftlicher Fachverlag

Innovationen mit Mikrowellen und Licht

Forschungsberichte aus dem Ferdinand-Braun-Institut, Leibniz-Institut für Höchstfrequenztechnik

Herausgeber: Prof. Dr. G. Tränkle, Prof. Dr.-Ing. W. Heinrich

Band 41: **Thi Nghiem Vu**
Development and analysis of diode laser ns-MOPA systems for high peak power application
ISBN: 978-3-7369-9480-5, 38,80 EUR, 138 Seiten

Band 42: **Christian Bansleben**
Differentieller Mikrowellen-Leistungsoszillator für die Realisierung ultrakompakter Plasmaquellen in Matrixanordnung
ISBN: 978-3-7369-9530-7, 34,90 EUR, 136 Seiten

Band 43: **Martin Winterfeldt**
Investigation of slow-axis beam quality degradation in high-power broad area diode lasers
ISBN: 978-3-7369-9733-2, 39,90 EUR, 158 Seiten

Band 44: **Jonathan Decker**
Investigation of monolithically integrated spectral stabilization in high-brightness broad area diode lasers
ISBN: 978-3-7369-9798-1, 49,50 EUR, 174 Seiten

Band 45: **Andreea Cristina Andrei**
Untersuchung und Optimierung robuster und hochlinearer rauscharmer Verstärker in GaN-Technologie
ISBN: 978-3-7369-9810-0, 39,90 EUR, 148 Seiten

Band 46: **Peng Luo**
GaN HEMT Modeling Including Trapping Effects Based on Chalmers Model and Pulsed S-Parameter Measurements
ISBN: 978-3-7369-9906-0, 48,00 EUR, 160 Seiten

Band 47: **Jörg Jeschke**
Entwicklung von optisch pumpbaren UVC-Lasern auf AlGaN-Basis Chalmers Model and Pulsed S-Parameter Measurements
ISBN: 978-3-7369-9918-3, 44,90 EUR, 176 Seiten

Band 48: **Nikolai Wolff**
Wideband GaN Microwave Power Amplifiers with Class-G Supply Modulation
ISBN: 978-3-7369-9931-2, 44,90 EUR, 170 Seiten

Band 49: **Carlo Frevert**
Optimization of broad-area GaAs diode lasers for high powers and high efficiences in the temperature range 200-220 K
ISBN: 978-3-7369-9944-2, 44,90 EUR, 174 Seiten

Band 50: **Bassem Arar**
GaAs-based components for photonic integrated circuits
ISBN: 978-3-7369-9976-3, 43,60 EUR, 152 Seiten

Cuvillier Verlag
Internationaler wissenschaftlicher Fachverlag

Innovationen mit Mikrowellen und Licht
Forschungsberichte aus dem Ferdinand-Braun-Institut, Leibniz-Institut für Höchstfrequenztechnik

Herausgeber: Prof. Dr. G. Tränkle, Prof. Dr.-Ing. W. Heinrich

Band 51: **Mahmoud Tawfieq**
Development and characterisation of a diode laser based tunable high-power MOPA system
ISBN: 978-3-7369-9983-1, 44,90 EUR, 170 Seiten

Band 52: **Sebastian Preis**
Hocheffiziente frequenzagile Mikrowellen-Leistungsverstärker auf Basis von Verbindungshalbleitern und Ferroelektrika
ISBN: 978-3-7369-7004-5, 54,00 EUR, 136 Seiten

Band 53: **Simon Fleischmann**
Materialaspekte der Hydridgasphasenepitaxie von Aluminiumgalliumnitrid
ISBN: 978-3-7369-7029-8, 41,80 EUR, 160 Seiten

Band 54: **Erhan Ersoy**
Optimierung von koplanaren GaN-MMIC-Leistungsverstärkern im X-Band
ISBN: 978-3-7369-7157-8, 39,90 EUR, 154 Seiten

Band 55: **Simon Rauch**
Lateral emission characteristics of high-power broad-area lasers subject to external optical feedback
ISBN: 978-3-7369-7168-5, 29,90 EUR, 112 Seiten

Band 56: **Frank Dittmar**
Untersuchung der Strahlgüte von brillanten Hochleistungs-Trapezlasern für den Wellenlängenbereich bei 808 nm
ISBN: 978-3-7369-7167-7, 44,90 EUR, 176 Seiten

Band 57: **Florian Hühn**
Flexibler Modulator und Digitalverstärker-MMIC für den energieeffizienten Betrieb einer digitalen Sendekette im GHz-Bereich
ISBN: 978-3-7369-7190-5, 29,90 EUR, 162 Seiten

Cuvillier Verlag
Internationaler wissenschaftlicher Fachverlag

www.ingramcontent.com/pod-product-compliance
Ingram Content Group UK Ltd.
Pitfield, Milton Keynes, MK11 3LW, UK
UKHW022000190726
13853UKWH00004B/1648